Sustainable Textiles: Production, Processing, Manufacturing & Chemistry

Series Editor

Subramanian Senthilkannan Muthu, Green Story Inc., Tung Chung, Hong Kong

This series aims to address all issues related to sustainability through the lifecycles of textiles from manufacturing to consumer behavior through sustainable disposal. Potential topics include but are not limited to: Environmental Footprints of Textile manufacturing; Environmental Life Cycle Assessment of Textile production; Environmental impact models of Textiles and Clothing Supply Chain; Clothing Supply Chain Sustainability; Carbon, energy and water footprints of textile products and in the clothing manufacturing chain; Functional life and reusability of textile products; Biodegradable textile products and the assessment of biodegradability; Waste management in textile industry; Pollution abatement in textile sector; Recycled textile materials and the evaluation of recycling; Consumer behavior in Sustainable Textiles; Eco-design in Clothing & Apparels; Sustainable polymers & fibers in Textiles; Sustainable waste water treatments in Textile manufacturing; Sustainable Textile Chemicals in Textile manufacturing. Innovative fibres, processes, methods and technologies for Sustainable textiles; Development of sustainable, eco-friendly textile products and processes; Environmental standards for textile industry; Modelling of environmental impacts of textile products; Green Chemistry, clean technology and their applications to textiles and clothing sector; Eco-production of Apparels, Energy and Water Efficient textiles. Sustainable Smart textiles & polymers, Sustainable Nano fibers and Textiles; Sustainable Innovations in Textile Chemistry & Manufacturing; Circular Economy, Advances in Sustainable Textiles Manufacturing; Sustainable Luxury & Craftsmanship; Zero Waste Textiles.

Subramanian Senthilkannan Muthu
Editor

Resource Mixes in Textiles

Case Studies

Springer

Editor
Subramanian Senthilkannan Muthu
Green Story
Hong Kong, Hong Kong

ISSN 2662-7108 ISSN 2662-7116 (electronic)
Sustainable Textiles: Production, Processing, Manufacturing & Chemistry
ISBN 978-3-032-10920-0 ISBN 978-3-032-10921-7 (eBook)
https://doi.org/10.1007/978-3-032-10921-7

This Springer imprint is published by the registered company Springer Nature Switzerland AG
The registered company address is: Gewerbestrasse 11, 6330 Cham, Switzerland

This book is dedicated to:
The lotus feet of my beloved Lord
Pazhaniandavar
My beloved late Father
My beloved Mother
My beloved Wife Karpagam and
Daughters—Anu and Karthika
My beloved Brother
Last but not least
To everyone working in various industrial
spaces to make the world, "SUSTAINABLE"

Contents

Introduction to the Book- Resource Mixes in Textile

Subramanian Senthilkannan Muthu

Abstract Nleashes a plethora of opportunities to reduce the carbon footprint. Biotechnology provides a variety of creative options for lowering industrial carbon footprints, such as carbon capture and utilisation of bioenergy and biofuels. In the evolving textile landscape, resource mix refers to the strategic integration of diverse raw materials and processing resources to reduce dependence on single-fiber systems, enhance sustainability, and improve performance efficiency. As the industry faces challenges such as resource depletion, waste generation, environmental pollution, and rising demand for eco-friendly products, adopting an effective resource mix becomes crucial for building a resilient and low-impact textile ecosystem. This book compiles emerging research and innovative approaches to sustainable material utilization—from ethnobotanical knowledge and agro-residue valorization to bio-based materials and circular recycling technologies. Covering themes such as carbon-neutral applications, utilization of cotton stalk residue, regenerative textile methodologies, natural and biodegradable fibers, and advancements in textile recycling, the book presents pathways for a climate-conscious and resource-efficient future. By highlighting opportunities and challenges, it encourages collaboration among researchers, manufacturers, and policymakers to unlock new possibilities toward a regenerative textile industry.

The second chapter explains from Agroforestry to Pharmaceuticals: Cross-Sectoral Ethnobotanical Approaches for Carbon–Neutral Industries. Ethnobotany plays a significant role in advancing carbon neutrality by combining traditional ecological knowledge with sustainable industrial practices. Indigenous communities, especially in the Western Himalaya, have used agroforestry, rotational farming, and sacred grove preservation to support carbon sequestration, conserve biodiversity, and manage natural resources sustainably. Their climate-resilient crops and wild plant-based food

---

S. S. Muthu (✉)
Green Story Incl, Toronto, Canada
e-mail: drsskannanmuthu@gmail.com

S. S. Muthu (ed.), *Resource Mixes in Textiles*, Sustainable Textiles: Production, Processing, Manufacturing & Chemistry, https://doi.org/10.1007/978-3-032-10921-7_1

systems also strengthen food security and environmental resilience. Replacing high-emission industrial goods with ethnobotanical alternatives—such as using plant-based dyes instead of synthetic ones, natural fibers for textiles, and biodegradable resins for adhesives—helps lower carbon footprints. Additionally, ethnoveterinary practices like using herbal feed additives can reduce methane emissions and encourage sustainable livestock management. Industries can also use plants to capture carbon by using medicinal and aromatic plants to purify the air, cleaning up pollution with special plants (phytoremediation), and adopting eco-friendly methods to extract plant-based materials while protecting the environment. Traditional healing systems, which include Ayurveda and Traditional Chinese Medicine (TCM), demonstrate sustainable medicinal plant use through regenerative agriculture, ethical harvesting, and eco-friendly processing techniques. To maximize the potential of ethnobotany in carbon neutrality, strong policy agendas and collaborations between policymakers, industries, and indigenous communities are essential. By integrating ethnobotanical principles into industrial strategies, industries can drive low-carbon development, circular economies, and sustainable innovations. The relationship between traditional knowledge and modern sustainability plays a significant role in developing carbon–neutral industries while at the same time preserving cultural heritage and biodiversity.

The third chapter is about the sustainable utilization of cotton stalk residue in textile resource mixes. The growing emphasis on sustainability in the textile industry has led to increased exploration of alternative fiber sources. Cotton stalk residue, a by-product of cotton cultivation, presents a viable and eco-friendly solution for fiber production. Traditionally considered agricultural waste, cotton stalks are rich in cellulose, making them a promising raw material for textile applications. This chapter explores into the chemical composition, fiber extraction techniques, and potential uses of cotton stalk fibers in sustainable textile resource mixes. Recent advancements in mechanical, chemical, and biotechnological processing have improved the feasibility of utilizing cotton stalks for nonwoven fabrics, blended textiles, and biocomposites. Additionally, the integration of cotton stalk fibers into industrial textiles such as packaging, geotextiles and paper-based materials is gaining momentum. The environmental benefits of repurposing cotton stalk residue include waste reduction, lower carbon emissions, and a significant contribution to the circular economy. Moreover, its economic advantages lie in reducing dependency on conventional fiber sources and providing cost-effective alternatives for manufacturers. Despite these benefits, challenges such as fiber quality variability, processing costs and market acceptance hinder widespread adoption. Addressing these issues through research, policy support, and technological innovations can enhance the commercial viability of cotton stalk fibers. This chapter aims to provide a comprehensive overview of the role of cotton stalk residue in textile resource mixes, highlighting its potential as a sustainable and economically feasible alternative in the global textile industry.

The fourth chapter is about the future of textile resource mix: Towards a regenerative industry. The textile industry is at a critical crossroads, where sustainability alone is no longer sufficient; a shift toward the regeneration of resources is essential. The major stakeholders in the textile sector need to examine the limitations of conventional textile production, including environmental degradation, over-reliance on finite resources, and challenges in waste management, and highlight emerging regenerative approaches that can redefine the industry. Innovative resource mixes in textiles can drive this transformation, which is the need of the hour. The future of textiles lies in regeneration and resource mix, moving beyond sustainability to actively restore ecosystems, enhance biodiversity, and minimise waste. This chapter examines the limitations of conventional fibre choices such as natural, synthetic, and blended, while highlighting regenerative alternatives that promote ecological balance. This chapter envisions a regenerative textile industry that restores rather than depletes, urging collaboration between fashion, agriculture, and biotechnology to create a net-positive impact.

The fifth chapter helps to understand the advancing sustainability in textiles: Natural Fibers, Bio-Based Materials, and Recycling Innovations. Nowadays, the textile industry is increasingly shifting towards sustainability to minimize its environmental footprint. This transformation focuses on adoption of eco-friendly materials such as natural fibers, bio-based textiles, and innovations in recycling technologies. Natural fibers like organic cotton, hemp, flax, jute, bamboo and others are gaining attention for their renewable properties and lower environmental impact compared to conventional textiles like polyester, non-organic cotton, rayon and other semi-synthetics. These fibers, however, come with challenges related to water usage and land requirements. Additionally, the emergence of bio-based and biodegradable textiles, derived from sources like algae, mushrooms, microbial cellulose, and fungal mycelium, presents a promising alternative to traditional fibres. These materials are valued for their environmental benefits, yet obstacles such as scalability, performance, and market acceptance need to be addressed for their broader application. Further, recycling plays a pivotal role in circular economy of textiles, contributing to reduction of waste and resource consumption. The recycling of materials such as cotton, polyester, and blended fabrics has seen advancements in technology, although challenges remain in optimizing the efficiency and effectiveness of these processes. The shift towards recycled and upcycled textiles is crucial for reducing the textile industry's overall environmental impact and establishing a more sustainable production cycle. Through this exploration, a comprehensive understanding of current sustainable practices in textile production is provided in this chapter, highlighting the opportunities, limitations, and future potential of these materials and technologies in shaping the future of the textile industry.

The sixth chapter is about the advancing sustainable practices in textiles: Exploring innovative resource utilization. The global textile industry is still working under a linear system known as take, make and dispose. This system is associated with environmental damage and different social problems. The current pattern leads to high use of natural resources, pollution in land and water bodies, and a huge amount of textile waste. Because of this, there is a strong need to move towards circular and sustainable methods to reduce these harmful effects. This chapter discusses how sustainable practices have gradually evolved in the textile industry and focus is given on innovative resource utilization as a key approach towards achieving sustainability. Under material innovation, several practices such as utilization of agricultural residues, development of textiles from fungal mycelium and marine algae, chemical recycling of post-consumer textile waste, and production of biodegradable polymers are discussed. Process-related innovations are highlighted with examples like resource-saving techniques in manufacturing including supercritical CO_2 dyeing and digital textile printing, along with detailed study of recycling technologies. In addition, this chapter also discusses the application of Life Cycle Assessment (LCA) as framework for evaluating sustainability and critically highlights major ongoing debates in the industry such as whether bio-based or recycled synthetic fibers are more sustainable. However, some knowledge gaps were also found in literature including lack of studies on integrated systems approach, limited understanding about social impacts of new technologies, and insufficient data available from developing countries perspective. For future direction, more emphasis should be given on interdisciplinary research to develop circular textile economy that is not only technologically feasible but also practically scalable and socially just.

I take this opportunity to thank all the contributors for their earnest efforts to bring out this book successfully. I am sure readers of this book will find it very useful.

With best wishes,

Dr. Subramanian Senthilkannan Muthu.

From Agroforestry to Pharmaceuticals: Cross-Sectoral Ethnobotanical Approaches for Carbon–Neutral Industries

Niharika Sharma and Radha

Abstract Ethnobotany plays a significant role in advancing carbon neutrality by combining traditional ecological knowledge with sustainable industrial practices. Indigenous communities, especially in the Western Himalaya, have used agroforestry, rotational farming, and sacred grove preservation to support carbon sequestration, conserve biodiversity, and manage natural resources sustainably. Their climate-resilient crops and wild plant-based food systems also strengthen food security and environmental resilience. Replacing high-emission industrial goods with ethnobotanical alternatives—such as using plant-based dyes instead of synthetic ones, natural fibers for textiles, and biodegradable resins for adhesives—helps lower carbon footprints. Additionally, ethnoveterinary practices like using herbal feed additives can reduce methane emissions and encourage sustainable livestock management. Industries can also use plants to capture carbon by using medicinal and aromatic plants to purify the air, cleaning up pollution with special plants (phytoremediation), and adopting eco-friendly methods to extract plant-based materials while protecting the environment. Traditional healing systems, which include Ayurveda and Traditional Chinese Medicine (TCM), demonstrate sustainable medicinal plant use through regenerative agriculture, ethical harvesting, and eco-friendly processing techniques. To maximize the potential of ethnobotany in carbon neutrality, strong policy agendas and collaborations between policymakers, industries, and indigenous communities are essential. By integrating ethnobotanical principles into industrial strategies, industries can drive low-carbon development, circular economies, and sustainable innovations. The relationship between traditional knowledge and modern sustainability plays a significant role in developing carbon–neutral industries while at the same time preserving cultural heritage and biodiversity.

Keywords Ethnobotany · Carbon neutrality · Sustainable industries · Indigenous knowledge

---

N. Sharma · Radha (✉)
School of Biological and Environmental Sciences, Shoolini University of Biotechnology and Management Sciences, Solan, India
e-mail: radhuchauhan7002@gmail.com

S. S. Muthu (ed.), *Resource Mixes in Textiles*, Sustainable Textiles: Production, Processing, Manufacturing & Chemistry, https://doi.org/10.1007/978-3-032-10921-7_2

1 Introduction

For centuries, people have depended on plants for many purposes, including food, medicine, shelter, decoration, construction, and clothing. Indigenous and local communities have vast knowledge of using these plants for various purposes, and using these plants with local traditional knowledge is known as traditional ethnobotanical knowledge. Nowadays, this knowledge is not well documented as the younger generation is not very interested in this traditional knowledge, but has been orally transmitted from one generation to another, and by doing this, the knowledge will one day disappear from the face of the earth [24]. However, to protect this knowledge, ethnobotanists have played an essential role in unravelling and documenting these plant-people interactions and have unlocked the knowledge by conducting various interviews and surveys. While there is not much information and records found about how humans used these plants in ancient times. Whereas, it was recently found that people in South Africa used the leaves of a plant called *Cryptocarya woodii* to make bedding about 77,000 years ago. Scientists were able to figure this out using advanced tools that study ancient plant DNA, called archaeogenetics [50]. This plant is still used by the people who live in the same area where it was discovered, and it was found that it helps keep mosquitoes away because it can kill their larvae. The recovery of starch granules from the surfaces of at least 105,000-year-old stone tools from Mozambique suggests that early humans consumed grass seeds [26]. Ethnobotany is a field that works at the intersection of plants and people at one end and science at the other end, therefore, ethnobotanists can act as a bridge between them [4]. The urgent challenge of climate change has elevated the search of carbon neutrality as a top priority for industries worldwide, requiring inventive and sustainable approaches to control greenhouse gas emissions and enhance the removal of carbon dioxide from the atmosphere [1, 51]. In response, there is a growing appreciation for the integration of ethnobotanical knowledge rooted in traditional plant use and management into contemporary industrial strategies for carbon neutrality. Ethnobotanical approaches provide distinctive avenues toward achieving carbon neutrality by utilizing the carbon-sequestering abilities of plants, encouraging the sustainable use of biomass, and aiding the restoration and preservation of natural ecosystems [7, 51]. Initiatives such as afforestation, producing bioenergy with carbon capture, and incorporating plant-based materials in manufacturing not only decrease dependence on fossil fuels but also promote circular resource flows in industrial systems [1]. The principle of carbon neutrality in biomass systems relies on the natural balance between carbon absorbed by plants during growth and the carbon released when biomass is utilized, emphasizing the need for sustainable management and short harvest cycles to maximize environmental benefits [7]. Utilizing both traditional ecological knowledge and modern technological advancements, ethnobotanical methods can enhance established approaches like renewable energy implementation, carbon capture and storage, and the transformation of waste

Table 1 Ethnobotanical contributions to eco-friendly alternatives and methane mitigation

Source	Study focus	Key findings	References
Plant-based alternatives	High-protein plant-based cheese	Developed using waxy starch, plant-protein isolate, and coconut oil; superior texture and meltability compared to commercial options	[16]
	Environmental impact of plant-based meat	Plant-based meat analogues reduce GHG emissions by 62% compared to beef burgers	[44]
	Life Cycle Assessment: Beef versus Plant Patties	Plant-based patties have significantly lower GHG emissions, water consumption, and land use	
	Sustainability of plant-based dairy	Offers 59–71% reduction in GHGs, less land and water use, maintains nutrition	[14]
Livestock methane mitigation	Feed-based methane mitigation	Low residual feed intake, diet quality, and concentrate-based rations reduce CH_4 per kg of milk	[8]
	Broad strategies for methane control	Multiple interventions (e.g., PSMs, feed additives, enzyme inhibitors); PSMs as a promising natural option	[3]
	Factors influencing CH_4 production	Feed type, genotype, microbial profiles, forage selection, and stocking rate influence methane; a systems-based approach is needed	Galati et al. (2024)

into valuable resources [1, 51]. Safeguarding and integrating ethnobotanical knowledge are crucial for preserving cultural traditions and developing innovative strategies to address contemporary environmental challenges. This chapter examines how indigenous knowledge supports carbon storage and sustainable resource management, while exploring its applications in various industries, including the development of eco-friendly products, methane reduction, and carbon capture. It also discusses the role of traditional medicine in promoting ethical resource use, reviews policies and industrial strategies for incorporating traditional ecological knowledge into carbon–neutral systems, and highlights the challenges, opportunities, and future directions for blending ancient wisdom with modern science to build sustainable, low-carbon industries (Table 1).

2 Indigenous Knowledge and Carbon Sequestration

Over the past thirty years, agroforestry has achieved worldwide recognition as a comprehensive approach to sustainable land management, appreciated for its productivity and beneficial effects on the environment. Its recent addition as a greenhouse gas mitigation measure under the Kyoto Protocol has brought greater attention to its role in biological carbon (C) sequestration. This potential arises from the concept that integrated systems like agroforestry make more efficient use of resources—such as nutrients, sunlight, and water—compared to monocultures, resulting in increased overall carbon capture and storage [29]. Capturing carbon from the atmosphere and storing it within terrestrial ecosystems is a recognized strategy for offsetting greenhouse gas emissions. Farmlands, especially when trees are incorporated and managed together with crops and livestock, offer considerable potential for carbon storage. Consequently, agroforestry is gaining acknowledgment not only for advancing sustainable agricultural practices but also for its significant contribution to addressing climate change challenges [2]. Agroforestry involves integrating trees into agricultural landscapes. This practice has proven beneficial for farmers in both developed and developing countries, contributing to enhanced soil health. While agroforestry provides distinct advantages, such as improved productivity and resilience, planting trees in degraded forest areas has also demonstrated effectiveness in fighting climate change by capturing and storing carbon [28]. India hosts thousands of sacred groves-forests preserved by local communities due to their cultural or spiritual importance. These sacred sites are often rich in biodiversity and play a crucial ecological role. Unlike government-managed protected areas, most sacred groves in India are overseen by community groups and are typically situated on private or communal land. This unique form of stewardship presents distinct challenges for future management and policy development concerning the conservation of these ecologically and culturally significant areas [31]. The rising population pressure, along with forest degradation and depletion, makes sustainable forest management a major challenge. In India, forests cover only about 23.8% of the country's total geographical area, placing it among the nations with low forest cover [22]. Many indigenous communities living in and near forests have long practiced the conservation of natural resources and biodiversity as part of their traditions. The idea of sacred groves dates back to ancient times, even mentioned in the Rig-Veda, where tree worship was widespread and deeply rooted in cultural practices. Particularly, the Buddha is said to have attained enlightenment while meditating under a banyan tree, further highlighting the historical significance of trees in spiritual and cultural life [30]. Traditional knowledge has been used for centuries by indigenous local communities to manage natural resources under local laws, customs, and traditions [13]. Across the globe, numerous traditional cultures hold nature in deep reverence, often regarding specific plants and animals as sacred. In the Western Ghats, sacred groves represent small remnants of ancient forests that are set aside in honour of local animistic deities. These groves serve as both spiritual sites and important reservoirs of biodiversity [49]. However, shifts in socio-economic factors and evolving land use practices have put many sacred groves at risk. These changes have led to reductions in their size, as well as alterations in their vegetation structure and the variety of species they support [11].

3 Ethnobotanical Contributions to Sustainable Industry Practices

Ethnobotany bridges scientific and traditional knowledge, enabling the identification of culturally significant and ecologically important plant species. This guides conservation priorities and supports community-driven conservation initiatives, such as the protection of sacred groves and the use of sustainable harvesting techniques [17]. In a study, researchers developed a high-protein (18% w/w) plant-based cheese alternative using waxy starch, plant-protein isolate, and coconut oil. This enhanced viscoelastic properties and meltability, outperforming commercial alternatives in texture and rheological properties [16]. Research indicates that plant-based meat analogues produce a substantially lower environmental footprint compared to beef burgers, resulting in a 62% reduction in greenhouse gas emissions [44]. Life cycle assessments reveal that plant-based patties consistently outperform beef burgers in reducing not only GHG emissions but also water consumption and land use [45]. Similarly, plant-based dairy alternatives have been shown to be both healthy and sustainable, offering reductions of 59–71% in greenhouse gas emissions, as well as lower land and water use relative to dairy milk, all while maintaining nutritional adequacy [14]. Ethnoveterinary practices present valuable approaches for reducing methane emissions in livestock, drawing on traditional knowledge and the use of natural remedies. Rooted in generations of experience, these methods can work alongside modern strategies to address methane emissions from ruminants, which are major sources of greenhouse gases. Research examining various ruminant management methods for enteric methane (CH_4 reduction in both extensive and intensive systems shows that selecting and breeding animals with low residual feed intake can significantly decrease CH_4 emissions. Higher productivity in dairy cows is linked with lower CH_4 emissions per kilogram of milk produced. Additionally, the review emphasizes the importance of diet composition, specifically, increasing concentrate intake and providing low-fiber, high-protein feeds are effective in lowering methane output. The choice of grain source and the proportion of concentrate in the diet are also critical. Overall, digestive physiology, diet quality, and selective breeding emerge as key strategies for methane mitigation in ruminant livestock [8]. Further, studies emphasize targeting ruminal fermentation, a key source of methane emissions, due to its large impact on global warming. A spectrum of approaches has been explored: vaccination, enzyme inhibitors, phage therapy, feed processing techniques, supplementation with fatty acids and organic acids, ionophores, microbial additives, non-ionic surfactants, halogenated methane analogs, sulphates, and defaunation, as well as dietary adjustments. Among these, plant secondary metabolites (PSMs) stand out as promising natural options, modifying rumen fermentation and reducing methane output. Managing methane from manure is also recognized as an effective supplementary strategy. Overall, reducing methane emissions supports climate goals while improving feed efficiency and livestock productivity, confirming the central role of rumen microbial management in animal nutrition [3]. Research has shown that methane emissions from ruminants are shaped by a range of factors, including the type of feed, rumen microbial communities, animal genotype, and the quality of the diet. Successful mitigation strategies incorporate nutritional adjustments, careful forage selection, and pasture management methods such as monitoring forage mass and adjusting stocking rates to optimize animal intake and decrease emissions. Important factors of methane production include dry matter intake, dietary composition, and interactions between microbes and the rumen environment. Notably, animals with higher feed efficiency do not always produce less methane, highlighting the importance of understanding the methanogenic characteristics of different genotypes. The study emphasizes the necessity for integrated, system-based research that brings together dietary, production, and management practices to craft effective and durable solutions for methane reduction (Galati et al. 2024).

4 Ethnobotany and Industrial Carbon Capture

Ethnobotanical research has revealed plant species with notable capacities for carbon sequestration, insights that frequently stem from traditional ecological knowledge. Indigenous and local communities manage landscapes in ways that promote biomass productivity and increased carbon storage—using agroforestry, mixed cropping systems, and conserving high-biomass plants. These time-tested practices can serve as models for large-scale industrial afforestation, reforestation, and land management initiatives focused on maximizing carbon sequestration [47]. In addition, studies assessing CO_2 sequestration technologies and their practical uses in industry highlight that post-combustion and oxy-fuel combustion processes are especially effective for carbon capture in the cement sector, with oxy-fuel combustion achieving higher efficiency. Carbon mineralization is shown to offer a promising approach for the permanent storage of CO_2, while transforming waste materials into usable resources. Integrated carbon capture and utilization (CCU) strategies further contribute to reducing emissions and replacing conventional raw inputs, thereby advancing industrial sustainability. Pyrolysis of lignocellulosic biomass is another efficient method for long-term carbon sequestration, producing biochar that stores carbon over extended periods. Additionally, microbial conversion using microalgae has shown success in transforming CO_2 into biofuels and valuable bioproducts. These findings highlight the promise of combining various CO_2 capture technologies to cut emissions, generate value-added products, and deliver both economic and environmental benefits [41]. A study evaluated the carbon emission mitigation potential of selected tree species around three industrial sites: weaving, brick kiln, and cosmetic manufacturing. The findings indicated that *Moringa oleifera* exhibited the highest total biomass at inside-industry locations, with values of 2.58, 0.56, and 4.57 Mg ha^{-1} across the respective sites. Among the species, *Syzygium cumini* had the greatest total carbon stock and sequestration within the weaving (2.82 and 10.32 Mg ha^{-1}) and brick kiln industries (3.78 and 13.5 Mg ha^{-1}), while *Eucalyptus camaldulensis* led in the cosmetic industry (7.83 and 28.70 Mg ha^{-1}). Outside the industrial areas, carbon stock and sequestration were generally lower. Nevertheless, top performers included *Dalbergia sissoo* for the weaving industry (0.97 and 3.54 Mg ha^{-1}), *Moringa oleifera* at the brick kiln site (1.26 and 4.63 Mg ha^{-1}), and *Vachellia nilotica* at the cosmetic industry site (2.51 and 9.19 Mg ha^{-1}). Belowground carbon data indicated that soil carbon stock was highest in the top 0–20 cm layer and decreased with depth. This trend was consistent across all industries and both inside and outside zones. Overall, the findings confirm that urban and industrial-adjacent tree species, especially *Syzygium cumini*, *Eucalyptus camaldulensis*, and *Moringa oleifera*, play a significant role in carbon sequestration and can be strategically utilized in nature-based industrial emission mitigation strategies [53]. Another study revealed that the industrial sector contributes 21% of global CO_2 emissions, necessitating targeted carbon capture strategies due to the sector's process diversity. Unlike power generation, industrial facilities typically feature numerous smaller emission sources with diverse CO_2 concentrations, making carbon capture efforts more complex and cost-sensitive. In chemical industries, however, high-purity CO_2 streams present more affordable opportunities for early deployment of carbon capture and storage (CCS). Despite this, CCS cost estimates vary widely between sectors, hindering broader, sector-wide adoption. Additionally, policy challenges remain significant—global market competition can disadvantage early CCS adopters unless measures like carbon pricing or trade adjustments are established to level the playing field [25]. Another study assessed the importance and economic viability of CCS technologies in achieving substantial reductions in greenhouse gas emissions, with a particular focus on their application in the cement industry. Estimated costs for coal-fired post-combustion capture (PCC), natural gas-fired PCC, and Cryogenic Carbon Capture (CCC) were examined, accounting for regional variations in fuel and capital expenses. These figures were incorporated into the MIT Economic Projection and Policy Analysis (EPPA) model to forecast long-term deployment of industrial CCS technologies under alternative policy scenarios. The analysis demonstrated that industrial CCS remains a vital mitigation tool, particularly where negative emission technologies are lacking, to meet strict climate objectives. The modelling showed that CCS allows for the continued use of energy-intensive industrial products while delivering substantial emission cuts across industries and regions. The developed methodology also supports more comprehensive CCS planning and its integration into global energy and economic modelling structures [18]. Additionally, another study investigated the CO_2 capture capability of industrial by-products and urban waste, targeting carbon neutrality goals by 2050. Using CO_2 fixation experiments with constant flow aeration, the research assessed a range of by-products especially those high in free calcium from industrial waste disposal—for their carbon sequestration capacities. The concept of 'Disposal Facility CCS' was introduced, and estimates showed that such facilities can offer significant CO_2 capture potential, in some cases outperforming natural forest sequestration. Using national data, the authors also projected the capture potential of industrial by-products across Japan, supporting the vision of a future urban environment designed for passive carbon capture. These findings highlight the viability of integrating industrial waste into broader climate mitigation strategies [23]. A different study highlights the growing relevance of indoor plants in enhancing indoor air quality (IAQ), especially in the post-COVID-19 context, where people spend most of their time in enclosed spaces. Evidence supports that species like *Ficus benjamina*, *Chlorophytum comosum*, and *Dracaena* sp. significantly reduce airborne pollutants, offering a natural and cost-effective method to combat sick-building syndrome and respiratory ailments. The pollutant removal efficiency varies with plant species and specific traits such as leaf area, thickness, photosynthetic rate, and involvement of plant parts (leaves, roots, stomata). Despite their proven benefits, standardized criteria for selecting optimal indoor plants and understanding their influence on microclimatic factors (humidity, temperature, ventilation) remain lacking. The findings highlight the need for controlled, real-time indoor environment simulations to better quantify the biophilic and air-purifying roles of houseplants [39]. Air phytoremediation is emerging as an effective and economical alternative to traditional air purification methods, which typically demand significant investment in specialized infrastructure and technical skills. Beyond their ability to remove pollutants, certain plants also play a role in sequestering CO_2 through photosynthesis, supporting efforts to mitigate greenhouse gas emissions (Table 2). This approach benefits from the combined action of plants and the diverse microbial communities that inhabit leaf surfaces, making it a sustainable solution both environmentally and economically. The success and efficiency of phytoremediation techniques largely depend on the unique characteristics of plant species and the diversity of microbes present on the leaves. Consequently, targeted research is essential to identify and develop optimal plant–microbe partnerships for effective urban air quality management [10].

5 Traditional Medicine and Sustainable Resource Management

The rapid depletion of natural resources—driven by urban expansion, global warming, and habitat loss—poses a significant threat to the sustainability of traditional medicine systems that depend on herbs, minerals, and animal-based ingredients. The use of animal products in particular often raises ethical and ecological concerns. Ayurveda, however, exemplifies a sustainable approach by promoting non-lethal, respectful harvesting methods, such as collecting milk, honey, or wool in ways that do not harm animals or disrupt ecosystems. These practices embody an ethical and environmentally sound model that can be revisited and adapted for modern needs. By integrating traditional wisdom, contemporary strategies for resource use in traditional medicine can be reframed to prioritize both conservation and the enduring practice of these systems [38]. In addition, the practice of ingredient substitution in traditional Tibetan medicine stands out as a vital adaptive strategy within changing socio-ecological conditions. This approach not only demonstrates sophisticated ecological understanding but also enables resilience in the face of resource scarcity and environmental changes. Through the lens of Resilience Theory, particularly concepts like adaptive cycles and panarchy, it becomes evident that traditional

Table 2 Ethnobotanical insights and technologies for industrial carbon capture

Aspect/Study focus	Key findings	Industrial and environmental implications	Reference
Ethnobotanical knowledge	Identifies plant species with high carbon sequestration; traditional practices (agroforestry, mixed cropping, high-biomass species)	Informs afforestation, reforestation, and land management for industrial-scale carbon sequestration	[47]
CO_2 capture technologies	Post-combustion and oxy-fuel combustion are effective in the cement industry; carbon mineralization for permanent storage; CCU strategies	Reduces emissions, enables waste-to-resource conversion, replaces raw materials, and supports sustainability	[41]
Biomass pyrolysis	Pyrolysis of lignocellulosic biomass produces biochar for long-term carbon storage	Efficient carbon sequestration method for industrial application	[41]
Microbial carbon conversion	Microalgae used to transform CO_2 into biofuels and bioproducts	Promising for converting industrial CO_2 emissions into valuable products	
Tree species for industrial sites	*Moringa oleifera*, *Syzygium cumini*, *Eucalyptus camaldulensis*, *Dalbergia sissoo*, *Vachellia nilotica* excels at carbon sequestration	Strategic use in urban/industrial zones for emission mitigation; species selection varies by industry type and location	[53]
Soil carbon storage	Highest soil carbon stock in top 0–20 cm layer, decreasing with depth	Soil management critical for maximizing belowground carbon sequestration	[53]
Industrial emissions and CCS	Industry responsible for 21% of global CO_2 emissions; multiple, diverse emission sources; high-purity CO_2 streams in some sectors	Targeted CCS strategies needed; cost and policy challenges for large-scale adoption	[25]
CCS cost-effectiveness	CCS critical for deep emission cuts (especially in cement); cost varies by region and technology; enables continued industrial output	Informs policy and planning for integrating CCS in global industrial sectors	[18]
Industrial by-products for CCS	Industrial by-products can fix CO_2; 'Disposal Facility CCS' concept; can outperform natural forests in some cases	Urban/industrial waste integration into climate mitigation; supports urban passive carbon capture vision	[23]

(continued)

Table 2 (continued)

Aspect/Study focus	Key findings	Industrial and environmental implications	Reference
Indoor plants and air quality	*Ficus benjamina, Chlorophytum comosum, Dracaena* sp. reduce indoor pollutants; efficiency varies by species and traits	Natural, cost-effective method for improving indoor air quality; need for standardization and further research	[39]
Air phytoremediation	Certain plants and phyllosphere microbes assimilate, degrade, or transform NO_2, SO_2, O_3, VOCs, PMs; also sequester CO_2 via photosynthesis	Low-cost, sustainable alternative for urban air quality management and greenhouse gas mitigation; requires targeted research	[10]

Tibetan medicine operates within a flexible, multi-scalar system that avoids rigidity traps. In contrast, the increasing industrialization and pharmaceutical standardization of traditional medicines impose inflexible frameworks that risk undermining sustainable resource use and cultural knowledge systems. Recognizing and institutionalizing the adaptive strategies inherent in traditional practices could support more resilient and sustainable medicinal production systems [20]. This study documented traditional ecological knowledge among residents of Chimalacatlán, focusing on local flora and fauna use and management. Using ethnographic surveys and chain-referral sampling, 48 individuals with extensive local knowledge were interviewed. Respondents identified 83 plant species and 44 animal species commonly used for food, medicine, fuel, ornamentation, and as pets. Key species identified for future sustainable management include *Odocoileus virginianus* (white-tailed deer) and *Ctenosaura pectinata* (Mexican spiny-tailed iguana) among animals, along with *Amphipterygium adstringens* and *Bursera linanoe* among plants. These findings highlight the significance of local knowledge in managing environmental resources and highlight the necessity for greater collaboration between communities and policymakers to ensure sustainable use and conservation efforts [21]. Traditional medicine systems, which are closely tied to natural ecosystems, exemplify the intrinsic relationship between environment, health, and culture. Deforestation poses a serious threat by diminishing biodiversity, undermining land rights, restricting access to resources, and jeopardizing cultural survival. Within this context, herbal medicine stands out as a sustainable practice that leverages locally available resources, thereby supporting ecological integrity, community well-being, and economic livelihoods. The study emphasizes the crucial role of herbal medicine in promoting sustainability, reinforcing the interconnectedness of nature, culture, and health [15]. The global rush for medicinal plants—which has intensified due to the COVID-19 pandemic, rising skepticism toward synthetic pharmaceuticals, and a renewed interest in natural remedies—has put considerable strain on wild plant populations. While these species offer immense therapeutic and economic value, unsustainable harvesting now threatens their survival and the ecosystems they support. Many medicinal plants are slow-growing, endemic, or keystone species, making them particularly vulnerable. Over-harvesting, habitat loss, and lack of local awareness compound the problem. A sustainable future centers on integrated approaches that blend scientific innovation, policy development, cultivation standards, and traditional knowledge. Community involvement and transdisciplinary collaboration are key to ensuring conservation while meeting global health demands [5]. Over the past two decades, the global interest in traditional medicine, particularly herbal remedies, has grown substantially. Both developed and developing nations are experiencing significant economic benefits from the expanding herbal market. However, despite the popularity of these products, reports highlight inconsistencies in quality. The production of herbal medicines is complex, with quality often compromised during harvesting, storage, and processing. Adopting good agricultural and harvesting practices is crucial for ensuring the safety and quality of medicinal plants. Harvesting the correct plant part at its optimal developmental stage, maintaining proper drying and storage conditions, and utilizing efficient processing techniques are key to producing consistent, high-quality herbal products. When these practices are followed, the safety, efficacy, and reliability of herbal medicines can be assured, boosting market potential and contributing to sustainable growth in the herbal industry [32]. Herbal medicines have been integral to healthcare since the beginning of human civilization and remain widely used today. Their therapeutic, pharmacological, and economic significance continues to gain recognition, though the extent of this varies across countries. Despite the growing use of medicinal plants, there is a lack of structured legislative controls, which has hindered their integration into modern medicine. One of the major barriers to this integration is the absence of solid scientific and clinical evidence regarding the efficacy and safety of herbal products. While herbal medicine research has expanded globally, many clinical researchers and ethics committees still express concerns about addressing the associated ethical issues. This article focuses on the clinical, ethical, and regulatory challenges surrounding herbal medicine, highlighting the need for a more structured approach to ensure their safe and effective use in healthcare [34]. Medicinal plants are central to India's Ayurvedic healthcare tradition. While only 10% of plant constituents are known, the demand for herbal drugs has surged, putting pressure on natural resources. Ayurvedic harvesting methods, validated by modern science, emphasize timing based on plant parts, seasons, and lunar cycles to maximize therapeutic efficacy. Environmental factors such as climate, altitude, and cultivation practices also influence plant metabolites, directly impacting the effectiveness of medicinal plants. Proper collection and processing techniques are essential for maintaining drug potency and therapeutic value [46]. Indigenous Indian medicine relies heavily on medicinal plants, yet only a fraction of the species has been thoroughly studied. Of the 300,000 known plant species, only a small number provide well-defined drugs, and only about 10% of their organic constituents are understood. The active principles in medicinal plants are made available through proper harvesting methods. Ancient Ayurvedic practices, as outlined by Acharya Susruta and Charaka, emphasize the importance of factors like plant part, soil conditions, and the desired therapeutic effects. Modern scientific data supports these methods,

Table 3 Sustainability challenges and adaptive strategies in traditional medicine systems

Category	Key findings	References
Sustainable use of animal resources in Ayurveda	Ayurveda promotes ethical, non-lethal use of animals (milk, honey, wool) that does not harm biodiversity or ecosystems	[38]
Adaptive strategies in Tibetan medicine	Ingredient substitution in Tibetan medicine is a flexible response to ecological change, reflecting resilience and deep ecological knowledge	[20]
Herbal medicine as a sustainability model	Herbal practices support ecological balance, local economies, and health, tied closely to land rights and biodiversity	[15]
Global demand and overharvesting of medicinal plants	COVID-19 and distrust of synthetics led to overharvesting; many medicinal species are slow-growing or keystone. Calls for integrated conservation methods	[5]
Quality concerns in herbal medicine production	Despite popularity, quality is often compromised during collection and processing; good harvesting and storage practices are essential	[32]
Regulatory and ethical challenges	Lack of legislation and clinical evidence hinders the integration of herbal medicine into modern systems; ethical concerns remain	[34]
Ayurvedic harvesting techniques	Traditional Ayurvedic harvesting is guided by timing, plant parts, and lunar cycles. Climate and altitude affect medicinal efficacy	[46]

highlighting the need for proper harvesting to preserve the plant's active compounds [36] (Table 3).

6 Policy and Industrial Integration

The circular economy is vital for sustainable development, aiming to decouple economic growth from resource depletion and environmental harm. To meet the EU Green Deal and net-zero objectives, the Visegrad (V4) countries, the Czech Republic, Hungary, Poland, and Slovakia, are implementing targeted strategies. A recent analysis examined the influence of the circular economy (CIR), green innovation (INV), renewable energy (REE), economic growth (GDP), and urbanisation (URB) on carbon emissions (CO_2) for the years 1990–2022 using CS-ARDL and quantile regression methods. The findings indicate that CIR, INV, and REE play a significant role in reducing CO_2 emissions, while GDP and URB contribute marginally to increased emissions [43]. The circular economy (CE) also delivers a transformative framework for sustainability within the tourism sector, fostering innovation and resource efficiency. In tourism and hospitality, research based on survey data from 180 professionals in Saudi Arabia and Egypt, analyzed using partial least squares structural equation modelling, reveals that CE encourages low-carbon behavior (LB), which in turn enhances environmental performance (EP). Moreover, eco-friendly behaviour (EB) further strengthens the connection between LB and EP, highlighting CE's influence on advancing sustainable industry practices and improving environmental results [52]. Additionally, policies promoting ethnobotanical approaches are critical for acknowledging indigenous knowledge, supporting sustainable resource use, and boosting biodiversity conservation. Collaboration among policymakers, industries, and indigenous communities enables better resource management, supports sustainable development, and ensures the inclusion of traditional knowledge within modern frameworks [24].

7 Challenges and Future Prospects

Ethnobotanical knowledge, while often grounded in empirical tradition, must undergo thorough scientific validation to establish the effectiveness of medicinal plants [42]. Around 80% of the world's population depends on plant-based medicines, highlighting the critical need to substantiate these practices with scientific evidence [37]. Historical records illustrate that numerous modern pharmaceuticals, including morphine and quinine, originated from ethnobotanical investigations, demonstrating the potential for discovering novel bioactive compounds [12, 37].

The commercial development of scientifically validated ethnobotanical products has the potential to create new markets for herbal medicines, generating benefits for both local communities and the broader global economy. However, ensuring quality control and standardization is essential for maintaining the safety and efficacy of these remedies, thereby fostering consumer confidence and wider acceptance in the marketplace [27]. High-amplitude ultrasonic processing has emerged as a promising technique for industrial processes in sectors such as food and beverage manufacturing. However, transitioning from laboratory to large-scale operations has been hindered by decreased ultrasonic intensity and trade-offs in product quality. The advent of barbell horn ultrasonic technology (BHUT) addresses this challenge, enabling high-amplitude ultrasonics at industrial scale and supporting scale-up factors exceeding 50 without compromising product quality. This development positions BHUT as a competitive alternative for commercial production [35]. In the context of industrial reliability, traditional testing methods encounter obstacles, particularly as comprehensive assessments of security and software reliability become increasingly vital. Furthermore, discrepancies between interconnect and gate delay reductions continue to challenge further scaling in industrial applications [40]. Looking ahead, ethnobotanical research is increasingly integrating technological innovations such as advanced databases, mobile applications, machine-learning techniques, Geographic Information Systems (GIS), remote sensing and big data analytics,. These tools greatly enhance data collection, management, and analysis, while assisting researchers in addressing issues of ethics and cultural sensitivity [6]. At the same time, the field is sharpening its emphasis on sustainable development and biodiversity conservation by integrating traditional indigenous wisdom with modern scientific methods. This fusion aims to facilitate the discovery of new drugs and refine conservation strategies, directly responding to pressing concerns about phytodiversity loss and ecological sustainability in a rapidly evolving global landscape [33].

8 Conclusion

Traditional ethnobotanical knowledge, together with indigenous communities, plays a vital role in advancing sustainable development goals and supporting a carbon–neutral future. TEK is intrinsically linked to sustainable ecosystem management, providing essential insights for biodiversity conservation and resource stewardship. Indigenous peoples, as custodians of this knowledge, are key to weaving these practices into broader strategies for sustainability. Such integration can greatly contribute to SDGs addressing poverty alleviation, health, responsible consumption, climate action, and the preservation of terrestrial life. TEK serves as a robust framework for managing resources sustainably, emphasizing ecosystem resilience and the conservation of biodiversity [9, 24]. For example, indigenous practices documented in regions like the Tehuacán Valley illustrate effective plant resource management, methods which can inform and enhance current sustainability initiatives [9]. Employing hybrid approaches that combine traditional wisdom with modern advancements can further industrial sustainability by introducing environmentally friendly technological solutions [9]. Indigenous communities possess a profound understanding of their local ecosystems, allowing for resource management that both sustains ecological balance and mitigates climate change [48]. Their practices are founded on respect and reciprocity with nature, naturally aligning with carbon–neutral objectives through the promotion of low-impact resource use and dedicated conservation efforts [48]. Collaborating with indigenous communities can enhance policy frameworks and the effectiveness of sustainability initiatives by incorporating a wider range of perspectives into decision-making [24]. While integrating traditional ethnobotanical knowledge (TEK) into industrial sustainability strategies provides substantial benefits, it also presents challenges. Past experiences of commercializing and misappropriating indigenous knowledge without due recognition or compensation have caused significant injustices. Therefore, ethical collaboration and the protection of indigenous intellectual property rights are crucial for building authentic partnerships and securing sustainable, equitable outcomes [48].

References

1. Al Khaffaf I, Tamimi A, Ahmed V (2024) Pathways to carbon neutrality: a review of strategies and technologies across sectors. Energies 17(23):6129

2. Albrecht A, Kandji ST (2003) Carbon sequestration in tropical agroforestry systems. Agr Ecosyst Environ 99(1–3):15–27
3. Alemneh T, Getabalew M (2019) Strategies to reduce methane emission in ruminants. Int J Ecol Ecosolution 6:16–22
4. Alexiades MN (2003) Ethnobotany in the third Millennium: expectations and unresolved issues. Delpinoa 45(1):15–28
5. Alum EU (2025) Sustainable harvesting of medicinal plants: balancing therapeutic benefits with environmental conservation. Agroecol Sustain Food Syst 49(3):380–385
6. Banisetti DK, Kosuri NP (2023) Ethnobotanical research in the digital age: harnessing technology for data collection and analysis. Int J Indig Herbs Drugs 1–6
7. Baral A (2014) Demystifying the carbon neutrality of biomass. The international council on clean transportation (ICCT)
8. Broucek J (2018) Methane abatement strategies based on genetics and dietary manipulation of ruminants: a review. Archivos de zootecnia 67(259):448–458
9. Casas A, Lira R, Torres I, Delgado A, Moreno-Calles AI, Rangel-Landa S, Campos N (2016) Ethnobotany for sustainable ecosystem management: a regional perspective in the Tehuacán Valley. Ethnobotany of Mexico: interactions of people and plants in Mesoamerica, pp 179–206
10. Chakravorty A, Raghavan V, Grace AN, Kim SY, Van Le Q (2022) Plants and microorganisms for phytoremediation of air. In: Current developments in biotechnology and bioengineering. Elsevier, pp 109–131
11. Chandrashekara UM, Sankar SJFE (1998) Ecology and management of sacred groves in Kerala, India. For Ecol Manag 112(1–2):165–177
12. Colombo ML, Dalfrà S, Scarpa B (2012) Scientific evidence of ethnobotanical and Mediterranean knowledge of food-and well-being plants. J Pharm Sci Res 4(1):1662
13. Correa CM (2001) Traditional knowledge and intellectual property. Geneva: The Quaker United Nations Office (QUNO), p 17. ISBN 2-9700323-0-9
14. Craig WJ, Messina V, Rowland I, Frankowska A, Bradbury J, Smetana S, Medici E (2023) Plant-based dairy alternatives contribute to a healthy and sustainable diet. Nutrients 15(15):3393
15. De Souza A, Ferreira P, Hüther C, de Jesus I, de Carvalho A, de Oliveira RPRF, Ferreira V (2024) The sustainable power of herbal medicine: bridging tradition and ecology, pp 1–12
16. Dobson S, Marangoni AG (2023) Methodology and development of a high-protein plant-based cheese alternative. Curr Res Food Sci 7:100632
17. Eduard C (2023) Conservation of biodiversity through ethnobotany and traditional knowledge. J Biodivers Bioprospecting Dev 9:26
18. Farrell J, Morris J, Kheshgi H, Thomann H, Paltsev S, Herzog H (2018) The role of industrial carbon capture and storage (CCS) in emission mitigation. In: 14th greenhouse gas control technologies conference Melbourne, pp 21–26
19. Galati RL, Silva LCRP, Fonseca ASR, da Silva Cabral L, Cerom MFSQ (2023) Challenges and strategies for reducing enteric methane emissions in sustainable animal production. Concilium 23(14):301–316
20. Glover DM (2021) Traditional medicines in a global economy: resource sustainability and resilience in the traditional Tibetan medical practice of ingredient substitution. Hum Ecol 49(1):33–42
21. Hernández-Tapia R, Valverde T, Aranda A, Martínez-Peralta C, Platas-Neri D (2018) Traditional knowledge as a basis for the development of a sustainable resource management program. Ethnobiol Lett 9(2):1–11
22. Kandari LS, Bisht VK, Bhardwaj M, Thakur AK (2014) Conservation and management of sacred groves, myths and beliefs of tribal communities: a case study from north-India. Environ Syst Res 3:1–10
23. Komine H, Saitou Y, Suzuki K (2024) Carbon dioxide capture potential of industrial by-products for post-carbon society. Environ Geotech 11(1):1–13
24. Kumar A, Kumar S, Komal, Ramchiary N, Singh P (2021) Role of traditional ethnobotanical knowledge and indigenous communities in achieving sustainable development goals. Sustainability 13(6):3062

25. Leeson D, Ramirez A, Mac Dowell N (2019) Carbon capture and storage from industrial sources. In: Bui M, Mac Dowell N (eds), Magnesium batteries: research and applications. Royal Society of Chemistry, pp 296–314
26. Mercader J (2009) Mozambican grass seed consumption during the middle stone age. Science 326(5960):1680–1683
27. Mukherjee PK, Harwansh RK, Bahadur S, Banerjee S, Kar A (2016) Evidence based validation of Indian traditional medicine–way forward. World J Tradit Chin Med 2(1):48–61
28. Murthy IK, Gupta M, Tomar S, Munsi M, Tiwari R, Hegde GT, Ravindranath NH (2013) Carbon sequestration potential of agroforestry systems in India. J Earth Sci Clim Chang 4(1):1–7
29. Nair PKR, Kumar MB, Nair VD (2009) Agroforestry as a strategy for carbon sequestration. J Plant Nutr Soil Sci 172(1):10–23
30. Negi CS (2005) Religion and biodiversity conservation: not a mere analogy. The International Journal of Biodiversity Science and Management 1(2):85–96
31. Ormsby AA (2011) The impacts of global and national policy on the management and conservation of sacred groves of India. Hum Ecol 39(6):783–793
32. Pandey AK, Savita R (2017) Harvesting and post-harvest processing of medicinal plants: problems and prospects. Pharma Innov J 6(12):229–235
33. Pandey AK, Tripathi YC (2017) Ethnobotany and its relevance in contemporary research. J Med Plants Stud 5(3):123–129
34. Parihar N, Barve K, Bhatt LK (2023) Clinical, ethical, and regulatory issues of botanicals. In: Phytopharmaceuticals and herbal drugs. Academic Press, pp 265–279
35. Peshkovsky AS (2017) From research to production: overcoming scale-up limitations of ultrasonic processing. In: Ultrasound: advances for food processing and preservation. Academic Press, pp 409–423
36. Pushpan R, Kumari H, Nishteswar K (2012) Influence of harvesting and processing methods on the active principles of medicinal plants. Int J Pharm Biol Arch 3:1283–1286
37. Qureshi R, Ghazanfar SA, Obied H, Vasileva V, Tariq MA (2016) Ethnobotany: a living science for alleviating human suffering. Evid-Based Complement Altern Med: eCAM 2016:9641692
38. Rastogi S, Kaphle K (2011) Sustainable traditional medicine: taking the inspirations from ancient veterinary science. Evid-Based Complement Altern Med 2011(1):151435
39. Ravindra K, Mor S (2022) Phytoremediation potential of indoor plants in reducing air pollutants. Front Sustain Cities 4:1039710
40. Seshan K (2018) Reliability issues: reliability imposed limits to scaling. In: Handbook of thin film deposition. William Andrew Publishing, pp 43–62
41. Shammi M, Akter B (2024) CO_2 sequestration for conventional utilization and industrial application. In: Decarbonization strategies and drivers to achieve carbon neutrality for sustainability. Elsevier, pp 299–319
42. Subba B, Srivastav C, Kandel RC (2016) Scientific validation of medicinal plants used by Yakkha community of Chanuwa VDC, Dhankuta, Nepal. Springerplus 5:1–14
43. Takyi KN, Gavurova B, Ofori C, Mikeska M, Sampene AK (2024) Assessing the role of circular economy and green innovation in mitigating carbon emissions in the Visegrad countries. Int J Renew Energy Dev 13(6):1149–1161
44. Tang M, Miri T, Soltani F, Onyeaka H, Al-Sharify ZT (2024) Plant-powered plates: a life cycle assessment of plant-based meats in tackling climate change. Preprints. https://doi.org/10.20944/preprints202404.1262.v1
45. Tang M, Miri T, Soltani F, Onyeaka H, Al-Sharify ZT (2024) Life cycle assessment of plant-based vs. beef burgers: a case study in the UK. Sustainability 16(11):4417
46. Tavhare SD, Nishteswar K (2014) Collection practices of medicinal plants-Vedic, Ayurvedic and modern perspectives. Int J Pharm Biol Sci 5:54–61
47. Teffo TK, Ramalepe P, Risenga I (2022) A short communication on the ethnobotany, phytochemistry, pharmacological evidence and ecosystem restoration potential of South African *Portulacaria afra*. J Med Plants Stud 10:112–115
48. Turner NJ, Cuerrier A, Joseph L (2022) Well grounded: indigenous peoples' knowledge, ethnobiology and sustainability. People Nat 4(3):627–651

49. Vipat A, Bharucha E (2014) Sacred groves: the consequence of traditional management. J Anthropol 2014(1):595314
50. Wadley L, Sievers C, Bamford M, Goldberg P, Berna F, Miller C (2011) Middle stone age bedding construction and settlement patterns at Sibudu, South Africa. Science 334(6061):1388–1391
51. Wang F, Harindintwali JD, Yuan Z, Wang M, Wang F, Li S, Chen JM (2021) Technologies and perspectives for achieving carbon neutrality. Innov 2(4):1–22
52. Zaki K (2024) The impact of circular economy on environmental performance in the tourism and hospitality industry: the role of low-carbon behavior and eco-friendly behavior. Int J Tour Hosp Stud 7(2):285–304
53. Zubair M, Yasin G, Qazlbash SK, Ul Haq A, Jamil A, Yaseen M, Guo W (2022) Carbon sequestration by native tree species around the industrial areas of Southern Punjab, Pakistan. Land 11(9):1577

Sustainable Utilization of Cotton Stalk Residue in Textile Resource Mixes

Samar Gogari and Radha

Abstract The growing emphasis on sustainability in the textile industry has led to increased exploration of alternative fiber sources. Cotton stalk residue, a by-product of cotton cultivation, presents a viable and eco-friendly solution for fiber production. Traditionally considered agricultural waste, cotton stalks are rich in cellulose, making them a promising raw material for textile applications. This chapter explores into the chemical composition, fiber extraction techniques, and potential uses of cotton stalk fibers in sustainable textile resource mixes. Recent advancements in mechanical, chemical, and biotechnological processing have improved the feasibility of utilizing cotton stalks for nonwoven fabrics, blended textiles, and bio-composites. Additionally, the integration of cotton stalk fibers into industrial textiles such as packaging, geotextiles and paper-based materials is gaining momentum. The environmental benefits of repurposing cotton stalk residue include waste reduction, lower carbon emissions, and a significant contribution to the circular economy. Moreover, its economic advantages lie in reducing dependency on conventional fiber sources and providing cost-effective alternatives for manufacturers. Despite these benefits, challenges such as fiber quality variability, processing costs and market acceptance hinder widespread adoption. Addressing these issues through research, policy support, and technological innovations can enhance the commercial viability of cotton stalk fibers. This chapter aims to provide a comprehensive overview of the role of cotton stalk residue in textile resource mixes, highlighting its potential as a sustainable and economically feasible alternative in the global textile industry.

Keywords Sustainability · Cotton stalk residue · Waste reduction · Bio-composites · Environment

---

S. Gogari · Radha (✉)
School of Biological and Environmental Sciences, Shoolini University of Biotechnology and Management Sciences, Solan, India
e-mail: radhuchauhan7002@gmail.com

S. S. Muthu (ed.), *Resource Mixes in Textiles*, Sustainable Textiles: Production, Processing, Manufacturing & Chemistry, https://doi.org/10.1007/978-3-032-10921-7_3

1 Introduction

1.1 Overview of Sustainability Challenges in the Textile Industry

The textile industry contributes substantially to climate change and overall global pollution thereby facing profound sustainability challenges which are multidimensional, spanning both environment and economy [1]. 8–10% of global carbon emissions are caused by this sector. These emissions surpass international aviation and shipping emissions, mainly because of extensive use of water and chemicals, energy-intensive production processes, as well as the release of microfibers into the environment during laundering [2]. Because of this new Fast fashion trend there is an exponential rise in the global fiber production since 2000 which has accelerated the textile waste generation. Most of this textile waste is either incinerated or ends up in landfills, releasing hazardous chemicals and greenhouse gases and exacerbating the environmental impacts [2, 3]. Currently, only 8% of total fiber production currently comes from recycled sources even after the introduction of recycled fibers. Moreover, recycling technologies because of their low scalability and affordability, where recycled polyester is primarily sourced from PET bottles instead of addressing textile-to-textile recycling directly [3]. Furthermore, the production of both synthetic fibers like polyester (derived from non-renewable petroleum) and natural fibers such as cotton and viscose carries substantial biodiversity, land use, and water consumption risks, along with chemical pollution from dyeing and finishing processes [3, 2]. Social sustainability remains a pressing concern, as the industry often relies on cheap labor in developing countries, raising issues around workers' rights and fair wages [1]. The complexity and global reach of textile supply chains further complicate traceability and transparency, hindering efforts to implement circular economy principles and corporate social responsibility initiatives [1]. As consumer awareness and regulatory pressures increase, the industry is being pushed to innovate and adopt sustainable business models, but significant barriers like high costs, technological limitations, and entrenched practices, still persist, emphasizing the urgent need for systemic transformation supported by robust research and policy intervention [1, 2]. The chapter explores the sustainable use of cotton stalk residue, an abundant agricultural by-product rich in cellulose, as an eco-friendly alternative fiber source for the textile industry. It covers the chemical and physical properties of cotton stalk fibers, advanced mechanical, chemical, and biotechnological extraction methods, and their applications in textiles, biocomposites, packaging, geotextiles, and pulp production. Emphasizing environmental benefits like waste reduction and carbon footprint savings, it also addresses economic advantages such as rural income generation and decreased reliance on conventional fibers. Despite promising potential, challenges related to fiber quality variability, processing costs, scalability, market acceptance, and policy gaps remain, requiring coordinated research, policy support, and consumer awareness to realize cotton stalks' role in advancing sustainable textile resource mixes and circular economy goals.

1.2 Need for Alternative and Eco-Friendly Fiber Sources

The need for alternative and eco-friendly fiber sources or resource mixes in the textile industry is driven by the unsustainable environmental impacts of conventional fibers, such as cotton and petroleum-based synthetics, which contribute to high water use, pesticide pollution, and persistent waste [4]. Recent research underscores the potential of plant-based natural fibers like abaca, bamboo, and hemp, which offer significant sustainability advantages due to their renewability, biodegradability, and lower resource requirements compared to traditional fibers [5]. Moreover, the exploration of agricultural residues and non-wood cellulose sources for man-made cellulosic fibers (MMCFs) provides a promising direction for reducing dependence on wood-derived pulp and synthetic fibers, as these alternatives can help mitigate deforestation and decrease the carbon footprint associated with textile production [6]. The development of biobased synthetic fibers, derived from biomass and even food residues, is gaining momentum, offering low carbon footprints and, in many cases, improved biodegradability, though ongoing research is needed to enhance their performance and scalability [4]. Hybrid fabrics, which blend natural and synthetic fibers, also contribute to sustainability by optimizing resource use, extending product lifespans, and enabling recyclability, thus supporting circular economy principles [7]. The global shift towards bio-based textiles is further propelled by stringent environmental regulations and rising consumer demand for eco-friendly products, leading to significant market growth and innovation in fiber development [8]. Studies have also highlighted innovative processing techniques, such as water retting for extracting fibers from unconventional sources like *Alpinia purpurata* pseudo stems, which can create new value chains while reducing agricultural waste [9]. As the textile industry continues to expand, the integration of diverse, eco-friendly fiber sources and resource mixes is essential to address environmental pressures, conserve resources, and build resilient supply chains for a sustainable future [6, 10].

1.3 Overview of Cotton Stalks as Agricultural Residue

Cotton stalks, the lignocellulosic residue left after cotton fiber harvest, represent a vast and underutilized agricultural by-product with significant potential as an alternative resource for the textile industry. Globally, annual production of cotton stalks exceeds 150 million tonnes, yet most are either burned or left to decompose, contributing to environmental pollution and resource wastage [11]. Recent research highlights that cotton stalks are rich in cellulose, making them a promising renewable feedstock for bio-based materials, including regenerated textile fibers, composites, and nanocellulose [12]. Advanced extraction and processing technologies, such as steam explosion and chemical treatments, have enabled the isolation of long, high-purity cellulose fibers from cotton stalks, with mechanical properties approaching those of conventional cotton fibers, thus supporting their suitability for textile applications [12,

13]. Life cycle assessments demonstrate that valorizing cotton stalks for fiber production can significantly reduce the environmental footprint of textiles by diverting waste from incineration or landfill, lowering greenhouse gas emissions, and decreasing reliance on wood or petroleum-based raw materials [12, 14]. Furthermore, the integration of cotton stalk fibers into resource mixes for textiles aligns with circular economy principles, promoting sustainable resource utilization and supporting rural economies by creating new value chains for farmers [11, 12]. Despite these advantages, challenges remain in standardizing collection, processing, and quality control, as well as in scaling up industrial applications to meet textile industry requirements [11]. Ongoing research is focused on optimizing pretreatment methods, improving fiber yield and quality, and developing cost-effective supply chains to fully realize the potential of cotton stalks as a sustainable alternative in textile resource mixes [11, 12]. As the demand for eco-friendly fibers grows and regulatory pressures intensify, cotton stalks stand out as a promising solution for advancing sustainability in the global textile sector.

1.4 Purpose and Scope of the Chapter

This chapter explores the sustainable utilization of cotton stalk residue as an alternative resource in textile production, which is an abundant, yet largely untouched agricultural by-product. With growing environmental concerns and the need to transition away from conventional and unsustainable fiber sources, the chapter highlights cotton stalks for their high cellulose content and potential applications across nonwoven fabrics, composites, geotextiles, and even pulp and paper industries. The primary aim is to present an in-depth analysis of cotton stalks as a viable component in eco-friendly textile resource mixes. The chapter systematically covers the chemical composition and physical characteristics of cotton stalk fibers, comparing them with other natural fibers, and evaluates various mechanical, chemical, and biotechnological methods for fiber extraction. It also delves into practical applications within textiles and beyond, outlining environmental and economic benefits while recognizing current challenges such as fiber variability, scaling issues, and limited market acceptance. Through this comprehensive approach, the chapter sets the foundation for understanding how cotton stalks can contribute to circular economy practices, reduce dependence on virgin and synthetic fibers, and support sustainable rural development. Overall, it aims to reinforce cotton stalks as an economically and environmentally sound solution to address the pressing sustainability challenges in the global textile industry.

2 Composition and Characteristics of Cotton Stalks

2.1 Agricultural Generation and Availability of Cotton Stalks

The agricultural generation and availability of cotton stalks are intrinsically linked to global cotton production patterns, with significant regional variations reflecting both agronomic practices and climatic conditions. Cotton is cultivated extensively across Asia, the Americas, and Africa, with the largest producers being India, China, the United States, Brazil, and Pakistan, each contributing substantially to the global output and, consequently, to the generation of cotton stalk residues [12, 15, 16]. In India, for the 2023–24 season, cotton pressing estimates reached 323.02 lakh bales (170 kg each), with the central zone (Gujarat, Maharashtra, Madhya Pradesh) accounting for the highest share at 198.5 lakh bales, followed by the south zone (Telangana, Andhra Pradesh, Karnataka, Tamil Nadu) at 72.75 lakh bales, and the north zone (Punjab, Haryana, Rajasthan) at 46 lakh bales [17]. Given that for every tonne of seed cotton produced, approximately 2–3 tonnes of cotton stalks are generated, the annual residue from India alone is immense, with estimates suggesting over 50 million tonnes of cotton stalks produced globally each year [12]. In China, the world's largest cotton producer, the Xinjiang region dominates national output, and the associated cotton stalk residue is similarly concentrated, supporting both local energy and material applications [12, 16]. The United States, with major production in the Southeast, Delta, and Southwest regions, also generates significant cotton stalk biomass, though much of it remains underutilized or is incorporated back into the soil. In Brazil, a record cotton crop in 2023/24 (14.56 million bales) has further increased the availability of cotton stalks, particularly in the Mato Grosso region, which leads national production [16]. African countries, such as Burkina Faso and Mali, contribute to the continent's cotton stalk output, though residue management practices vary widely and are often influenced by local needs for fodder and fuel [18]. Despite the vast quantities produced, the economic utilization of cotton stalks remains limited in many regions, with burning and field decomposition still common, underscoring the need for improved residue management and valorization strategies to harness this abundant agricultural by-product for sustainable applications.

2.2 Chemical Composition

Cotton stalks, as a lignocellulosic agricultural residue, exhibit a complex chemical composition that makes them a valuable resource for various industrial applications, including textiles and bio-based materials. The primary constituent of cotton stalks is cellulose, typically comprising around 40–45% of the dry weight, which imparts structural rigidity and is a key target for valorization in fiber and bioproduct applications [19, 20]. Hemicellulose content generally ranges from 20–25%, providing a matrix that binds cellulose fibers and contributes to the flexibility and reactivity of the biomass [20]. Lignin, accounting for approximately 20–25% of the composition, acts as a binding agent and imparts resistance to microbial attack and chemical degradation, though it also poses challenges for processing due to its recalcitrance [20, 21]. The moisture content of cotton stalks is reported to be around 8–10%, influencing storage stability and processing efficiency, as excessive moisture can promote microbial spoilage and complicate thermal conversion processes [22]. Ash content, a measure of inorganic mineral residue after combustion, typically falls between 5–8%, with higher values potentially impacting the efficiency and quality of downstream applications such as biochar production or fiber extraction [19, 22]. Elemental analysis further reveals a predominance of carbon (about 65%), with minor proportions of hydrogen, oxygen, and nitrogen, reflecting the organic nature of the biomass and its suitability for conversion to energy or material products [22]. Advanced spectroscopic analyses, such as FTIR (Fourier Transform Infrared Spectroscopy) and XRD (X-Ray Diffraction), confirm the presence of crystalline cellulose and characteristic functional groups associated with lignin and hemicellulose, while chemical and microbial treatments can selectively reduce lignin and hemicellulose content, enhancing the accessibility and purity of cellulose for further processing [19, 23]. These compositional features underscore the versatility of cotton stalks as a renewable feedstock, while also highlighting the need for tailored pretreatment and processing strategies to maximize their value in sustainable material applications.

2.3 *Physical Characteristics of Stalk Fibers*

Cotton stalk fibers exhibit a range of physical characteristics that position them as promising materials for textile and composite applications, with their properties influenced by both their anatomical origin within the stalk and the extraction processes employed. The fiber length of cotton stalk fibers typically varies depending on the section of the stalk, with studies reporting that fibers from the middle section possess the greatest length and uniformity, which is advantageous for spinning and composite reinforcement, while root section fibers tend to be shorter and less uniform [24]. Tensile strength also shows significant variation by stalk section, with middle section fibers demonstrating the highest values (up to 56.3 cN/tex) compared to top Sect. (39.79 cN/tex) and root section fibers (2.21 cN/tex), reflecting differences in cell wall structure and chemical composition across the stalk [24]. The elasticity of cotton stalk fibers, as measured by elongation at break and modulus of elasticity, is closely linked to their lignin and cellulose content, with higher lignin-to-cellulose ratios imparting greater toughness and resistance to deformation, especially in the phloem compared to xylem tissues [25]. The structure of cotton stalk fibers is characterized by a composite arrangement of cellulose microfibrils embedded in a matrix of hemicellulose and lignin, which together determine the fiber's mechanical performance and response to stress [25]. Surface morphology studies indicate that untreated fibers are coarse and covered with residual gums, but chemical or enzymatic treatments can effectively remove these surface impurities, resulting in smoother fibers with enhanced wettability and improved interfacial bonding in composites [23, 26]. Additionally, the physical properties such as diameter, linear density, and moisture regain are significantly affected by both the anatomical location and the extraction technique, with middle and top section fibers generally exhibiting superior mechanical and physical properties suitable for high-value applications [24]. These findings underscore the importance of targeted extraction and processing strategies to harness the full potential of cotton stalk fibers for sustainable material development.

2.4 Comparative Analysis with Other Natural Fibers

Cotton stalk fibers, derived from the bark of cotton plant residues, present a unique profile when compared with other natural fibers such as jute, flax, hemp, and traditional cotton lint, offering both advantages and certain limitations for textile and composite applications. Chemically, cotton stalk fibers are rich in cellulose, with content reaching up to 82.1% after optimized treatments, which is comparable to or even higher than that of many other lignocellulosic fibers, and they possess a moderate amount of lignin and hemicellulose, contributing to their mechanical robustness and durability [13, 27]. Mechanically, the tensile strength and modulus of cotton stalk fibers are reported to be close to those of cotton fibers and superior to other agricultural byproduct fibers such as rice and wheat straw, making them attractive for reinforcement in composites and potentially suitable for textile spinning when processed to reduce fineness and improve aspect ratio [13, 27]. While the single cells of cotton stalk fibers are shorter and finer than those of cotton lint, advanced extraction methods, such as steam explosion combined with alkali and peroxide treatments can yield long, clean fibers with a fineness of 27 dtex and an aspect ratio of 1,150, approaching the requirements for textile-grade fibers [13, 26]. In comparison, fibers like jute and flax are known for their high tensile strength and long fiber length, which make them staples in the natural fiber composite industry, but cotton stalk fibers, with their high cellulose content and improved processing methods, are closing this performance gap and offer the added benefit of utilizing an abundant agricultural residue [13, 28]. Furthermore, natural fiber composites reinforced with cotton stalk fibers have demonstrated mechanical properties similar to those of jute-reinforced composites, highlighting their potential as a sustainable alternative [27, 28]. Overall, the valorization of cotton stalk fibers not only supports waste reduction and circularity in cotton agriculture but also provides a competitive, eco-friendly alternative to both conventional natural fibers and synthetic reinforcements in diverse material applications. Table 1 compares cotton stalks with other natural fibers with respect to various parameters.

Table 1 Comparison of cotton stalk fibers with other natural fibers

Parameters	Cotton stalk fibers	Jute/Flax/Hemp/Cotton lint	References
Source	Bark of cotton plant residues	Stem (jute, flax, hemp), seed hair (cotton lint)	[27]
Cellulose content	Up to 82.1% after optimized treatments	Comparable or slightly lower (varies by species)	[13]
Lignin and hemicellulose	Moderate—contributes to strength and durability	Varies by fiber type; flax and hemp generally lower lignin	[27]
Tensile strength and modulus	Comparable to cotton fibers; superior to straw fibers	Higher in jute/flax; cotton lint high in strength but finer	[13, 27]
Fiber length and fineness	Short and fine as raw; can reach 27 dtex, aspect ratio ~ 1,150 with treatment	Jute and flax have naturally long fibers; cotton lint has finer, longer single cells	[13, 26]
Processing techniques	Requires treatments like steam explosion, alkali, peroxide for fiber quality improvement	Jute/flax: mechanical retting; Cotton lint: ginning	[13, 26]
Composite reinforcement use	Mechanical properties similar to jute composites; promising for bio-composite applications	Jute/flax widely used in bio-composites	[28, 27]
Textile application potential	Suitable for spinning after refining; current developments closing gap with traditional textile fibers	Cotton lint: ideal for textiles; Jute/flax/hemp: coarse textiles or technical fabrics	[13, 26]
Sustainability	High—agricultural residue valorization, circular economy support	High—renewable, but often primary crops	[28]
Overall advantages	Abundant byproduct, high cellulose content, good mechanical performance, potential eco-friendly alternative	Proven strength, longer fibers; well-established in composites and textiles	[13, 28]

3 Fiber Extraction Techniques

3.1 Mechanical Methods

Mechanical extraction techniques for cotton stalk fibers, including decortication, crushing, carding, and milling, play a crucial role in separating fiber bundles from the woody core and bark, enabling their use in textiles and composites. Decortication is the primary mechanical process that applies shearing and tearing forces to weaken and separate the fiber bundles from the lignocellulosic matrix of the stalk, effectively removing shives and woody material. This method is widely used due to its efficiency in bulk fiber recovery, though it often results in fibers with variable length and some degree of breakage, which can affect the quality for spinning applications [29]. Following decortication, crushing further breaks down the stalk material, facilitating the release of finer fiber bundles and reducing the size of residual shives, which aids in subsequent cleaning and processing steps [29]. The carding process then aligns and cleans the fibers by disentangling and separating them into more uniform strands, improving fiber orientation and preparing them for spinning or composite reinforcement,this step also removes dust and short fibers, enhancing fiber quality [13]. Finally, milling can be applied to reduce fiber size and refine fiber morphology, though excessive milling may lead to fiber damage and reduced mechanical properties [29]. While mechanical methods are advantageous for their lower chemical usage and faster processing times, they often yield fibers with lower purity and shorter lengths compared to chemical or combined treatments, limiting their direct application in high-quality textiles [29, 30]. Recent studies have explored combining mechanical extraction with mild chemical treatments, such as alkali or peroxide, to improve fiber fineness, tensile strength, and surface cleanliness, thereby enhancing the fibers' suitability for textile and composite uses [13, 26]. For example, steam explosion pretreatment followed by mechanical decortication has shown promise in producing fibers with improved mechanical properties and higher cellulose content, indicating that integrating mechanical and chemical methods can optimize fiber extraction from cotton stalks for sustainable material applications [13, 30].

3.2 *Chemical Processes*

Chemical extraction processes are central to obtaining high-quality fibers from cotton stalks, with alkali treatment, bleaching, and solvent extractions being the most widely used and effective methods. Alkali treatment, typically employing sodium hydroxide (NaOH), is fundamental for breaking down the lignin and hemicellulose matrix that binds cellulose microfibrils, thereby facilitating the liberation of cellulose fibers from the stalk's bark; this process not only increases the yield but also enhances fiber purity and mechanical properties, as evidenced by studies demonstrating that alkali-treated cotton stalk fibers exhibit improved tensile strength and surface cleanliness, making them suitable for composite and textile applications [31]. Following alkali treatment, bleaching is often carried out using agents such as hydrogen peroxide (H_2O_2) to further remove residual lignin, pigments, and other non-cellulosic impurities,this step is critical for achieving the whiteness, brightness, and softness required for textile-grade fibers, with research showing that combined alkali and peroxide treatments can yield fibers with cellulose content as high as 82.1% and fineness values approaching those of conventional textile fibers [13]. Solvent extractions utilizing organic solvents, ionic liquids, or green solvent systems, offer an alternative or complementary approach for selective removal of lignin and hemicellulose, enabling the extraction of cellulose with tailored physicochemical properties,for instance, microwave-assisted extraction using glycerin solvents has been shown to efficiently produce high-quality natural fibers from cotton stalk bark, with enhanced crystallinity and mechanical performance [26]. The integration of these chemical processes, and their optimization through parameters such as concentration, temperature, and reaction time, is crucial for maximizing fiber yield, purity, and performance, as well as for minimizing environmental impact. Recent advances also highlight the potential for combining chemical and mechanical methods, such as steam explosion followed by alkali and peroxide treatments, to achieve the most efficient extraction of long, clean, and strong cellulose fibers from cotton stalks, supporting their use as sustainable alternatives in the textile industry.

3.3 Biotechnological and Enzymatic Methods

Biotechnological and enzymatic methods, particularly microbial and enzymatic retting, have emerged as eco-friendly advancements in the extraction of fibers from cotton stalks, offering significant advantages over conventional chemical techniques by reducing environmental impact and improving fiber quality. Microbial retting utilizes specific bacteria or fungi to selectively degrade pectins, hemicellulose, and lignin that bind cellulose fibers, thereby facilitating the gentle separation of fibers from the stalk matrix without harsh chemicals; for instance, inoculation with *Bacillus* species under controlled fermentation conditions has been shown to efficiently release fiber bundles while preserving their mechanical integrity and minimizing surface damage [26]. Enzymatic retting, often employing commercial preparations of pectinases, xylanases, and ligninases, further enhances the process by targeting specific non-cellulosic components, resulting in cleaner, softer, and more uniform fibers with improved wettability and spinnability, which are highly desirable for textile applications [12]. Recent research highlights that these biotechnological approaches can be precisely controlled to optimize fiber yield, fineness, and tensile properties, and they are compatible with subsequent mechanical processing steps such as carding and spinning [26]. The adoption of microbial and enzymatic retting also aligns with sustainable development goals by reducing the consumption of water, energy, and hazardous chemicals, and by generating less toxic effluent compared to alkali or bleaching treatments [12]. Moreover, advances like microbial retting with mild thermal or physical pretreatments in process engineering, have further improved extraction efficiency and fiber quality, making these methods increasingly viable for industrial-scale applications. As the textile industry seeks greener alternatives for fiber sourcing, biotechnological and enzymatic extraction from cotton stalks stands out as a promising, scalable solution that supports circular economy principles and the valorization of agricultural residues [12].

3.4 Comparison of Methods

The comparison of fiber extraction methods from cotton stalks like mechanical, chemical, and biotechnological, reveals distinct differences in efficiency, environmental impact, and scalability that are critical for sustainable material sourcing. Mechanical methods, such as decortication and milling, are straightforward and avoid chemical use, offering rapid processing and relatively low environmental impact; however, they often yield fibers with lower purity, variable length, and higher levels of non-cellulosic impurities, which can limit their direct application in high-quality textiles and require additional cleaning steps [30, 32]. Chemical extraction, particularly alkali treatment and bleaching, is highly efficient in removing lignin and hemicellulose, producing fibers with high cellulose content, improved fineness, and enhanced mechanical properties suitable for textile and composite applications [26, 31]. Nevertheless, these processes consume significant amounts of water and chemicals, generate alkaline effluents, and require careful management to mitigate environmental risks, though advances in process optimization and effluent treatment are improving their sustainability profile [31, 32]. Biotechnological and enzymatic methods, including microbial and enzymatic retting, stand out for their eco-friendly credentials, as they use specific microorganisms or enzymes to selectively degrade non-cellulosic components, resulting in cleaner fibers with minimal chemical input and lower water and energy consumption [30]. These methods also produce less toxic effluent and are increasingly compatible with industrial-scale operations due to advances in process control and enzyme production, though their extraction rates and processing times can be longer than those of chemical methods [26, 30]. In terms of scalability, mechanical and chemical methods are already established at industrial levels, while biotechnological processes are rapidly advancing toward commercial viability, especially as environmental regulations tighten and demand for green processing grows [26, 30]. Ultimately, the choice of extraction method depends on the desired fiber quality, environmental considerations, and economic factors, with integrated or hybrid approaches, combining mechanical, chemical, and biotechnological steps, emerging as optimal solutions for balancing efficiency, sustainability, and scalability in cotton stalk fiber extraction. Table 2 gives a comparative overview of various extraction methods for cotton stalk fibers.

4 Applications in Sustainable Textiles

4.1 Nonwoven Fabrics and Insulation Materials

Nonwoven fabrics produced from natural and reclaimed fibers are increasingly recognized as effective and sustainable materials for acoustic panels and thermal insulation, offering a viable alternative to synthetic and mineral-based solutions. The porous structure of nonwoven composites, often comprising blends of fibers such as cotton,

Table 2 Comparison of cotton stalk fiber extraction methods

Aspects	Mechanical methods	Chemical methods	Biotechnological methods	References
Process type	Decortication, milling	Alkali treatment, bleaching	Microbial and enzymatic retting	[26, 30–32]
Efficiency	Moderate—fast processing but lower fiber purity	High—effective removal of lignin and hemicellulose	Moderate—clean fibers but slower processing	[26, 30]
Fiber quality	Lower purity, variable length, higher impurities	High cellulose content, improved fineness and mechanical strength	Cleaner fibers with minimal chemical residue	[26, 31]
Environ-mental impact	Low—no chemicals used, but may require post-cleaning	High—uses chemicals and water, produces effluents	Low—low water and chemical use, less toxic effluent	[30, 32]
Resource consump-tion	Low energy and chemical usage	High water and chemical usage	Lower water, energy, and chemical consumption	[30, 32]
Scalability	High—already used at industrial scale	High—industrially established	Moderate to high—advancing toward industrial use	[26, 30]
Limitations	Fiber quality issues; may need further processing	Environmental pollution risks; needs effluent treatment	Slower extraction rates; limited by microbial/enzyme optimization	[30, 32]
Sustain-ability potential	Good if integrated with green post-processing	Moderate; improving with better chemical recovery and effluent treatment	High; strong eco-friendly potential, increasingly scalable	[26, 30]

pigeon pea stalk, or flax, creates interconnected networks of macropores and micropores, which are highly effective at dissipating sound energy and providing sound absorption across a wide frequency range [33, 34]. Recent studies demonstrate that composites made from waste cotton and pigeon pea stalk fibers, especially when optimized with compatibilizers, show improved dispersion of fibers in the matrix, resulting in enhanced acoustic absorption and increased thermal stability, making them suitable for use in aerospace, automotive, and building applications where both noise reduction and thermal management are critical [33, 35]. Nonwoven composites with a surface layer of cotton fiber exhibit higher fabric density, which translates to superior sound insulation compared to glass fiber or activated carbon fiber composites, as greater density and fiber packing increase the sound transmission loss and absorption coefficients [36, 37]. Additionally, the carbonization and activation of cotton nonwovens can further enhance their acoustic properties by increasing surface area and creating additional microporosity, which allows for greater adsorption and dissipation of sound waves [36]. Natural fiber-based nonwovens not only match or exceed the sound absorption performance of traditional materials like mineral wool and polyurethane foam but also offer advantages of biodegradability, lower environmental impact, and reduced health risks associated with fiber inhalation [38]. The thermal insulation performance of these materials is similarly promising, as the fibrous structure traps air, reducing heat transfer and providing effective insulation. Overall, the integration of agricultural residues and reclaimed fibers in nonwoven composites supports the development of multifunctional, eco-friendly materials for advanced acoustic and thermal insulation applications.

4.2 Blended Textiles

Blended textiles incorporating cotton stalk fibers with conventional fibers such as cotton, polyester, wool, and jute are gaining attention for their potential to enhance sustainability and functional properties in textile applications. When cotton stalk bark fibers are blended with cotton, the resulting yarns exhibit improved evenness, reduced hairiness, and favorable tensile properties, with fabrics demonstrating higher bursting strength and a softer, smoother handle compared to traditional jute/cotton blends, while also maintaining good dye sorption characteristics [32]. The mechanical properties of cotton stalk fibers, especially after advanced extraction treatments like steam explosion combined with alkali and peroxide, approach those of cotton fibers, with fineness values as low as 27 dtex and aspect ratios suitable for spinning, making them viable for integration into cotton-rich yarns for apparel and home textiles [13]. Blending cotton stalk fibers with polyester or polypropylene in composite materials has been shown to improve mechanical strength, thermal stability, and crystallinity, particularly when the fibers are chemically treated to enhance compatibility and dispersion within the polymer matrix, though surface roughness and fiber-matrix interactions must be carefully managed for optimal impact resistance [23]. Cotton stalk fibers also compare favorably to jute in composite reinforcement, with similar flexural, tensile, and impact resistance properties, supporting their use in technical textiles and nonwovens [27]. While direct blending with wool is less documented, the cellulose-rich composition and moderate modulus of cotton stalk fibers suggest potential for hybrid yarns that balance strength, elasticity, and environmental performance. Overall, the integration of cotton stalk fibers with established textile fibers not only valorizes agricultural residues and reduces environmental impact but also enables the development of innovative, high-performance, and eco-friendly textile products suitable for diverse end uses.

4.3 Biocomposites and Reinforced Textiles

Cotton stalks are increasingly recognized as a valuable resource for the development of biocomposites and reinforced textiles with significant applications in the automotive, construction, and sports industries due to their renewable, biodegradable nature and favorable mechanical properties [39–41]. In the automotive sector, biocomposites made from cotton stalk fibers, often combined with biodegradable polymers or bio-based matrices, are used for manufacturing interior and exterior panels, structural components, and trim parts, offering a sustainable alternative to petroleum-based materials while enabling weight reduction, cost efficiency, and design flexibility [40–42]. The high strength-to-weight ratio of these composites contributes to improved fuel efficiency and reduced carbon emissions, addressing key environmental challenges facing the industry, and their biodegradability supports end-of-life disposability without polluting the environment [40–43]. In construction, cotton stalk-based biocomposites are utilized for producing lightweight panels, insulation boards, and structural elements, leveraging their good thermal and acoustic insulation properties, resistance to corrosion, and capacity to be molded into complex shapes [39, 44]. These materials not only lower the carbon footprint of buildings but also provide a sustainable solution for managing agricultural residues, aligning with circular economy principles [39]. In the realm of sports equipment, the flexibility, impact resistance, and customizability of cotton stalk-reinforced composites make them suitable for applications such as protective gear, racquet frames, and lightweight boards, where performance and sustainability are both critical [41]. Ongoing research and industrial case studies confirm that the integration of cotton stalk fibers into biocomposite formulations can match or surpass the mechanical and functional properties of traditional synthetic composites, while also reducing dependency on non-renewable resources and promoting environmental stewardship across multiple sectors.

5 Industrial Utilization Beyond Textiles

5.1 Packaging and Molded Materials

Cotton stalks, an abundant agricultural residue, are increasingly being recognized as a promising raw material for the production of biodegradable packaging and molded materials, offering a sustainable alternative to conventional plastics and wood-based products. Recent research demonstrates that cotton stalk cellulose can be effectively utilized as a substrate for microbial synthesis of lactic acid, which serves as a precursor for polylactic acid (PLA), a widely used biodegradable bioplastic [45]. Through the action of specific microbial strains, such as *Lactococcus cremoris*, high yields of optically pure lactic acid can be achieved from hydrolyzed cotton stalks, enabling the direct synthesis of PLA with desirable properties for food, pharmaceutical, and general packaging applications. PLA derived from cotton stalks is not only biocompatible and biodegradable but also supports a low-carbon and circular bioeconomy by valorizing agricultural waste and reducing reliance on fossil resources [45]. In addition to PLA, cotton stalk powder has been successfully incorporated into wood–plastic composites (WPCs), improving the mechanical and thermal performance of molded materials and broadening their application in packaging and consumer goods [44]. Comparative life cycle assessments confirm that utilizing cotton stalks for biobased materials, including packaging, significantly reduces greenhouse gas emissions and environmental impact compared to traditional incineration or landfill disposal, supporting broader ecological and economic goals [14]. The valorization of cotton stalks into biodegradable packaging materials not only addresses the challenge of agricultural waste management but also aligns with global sustainability targets by providing cost-effective, renewable, and environmentally friendly alternatives to synthetic plastics [12, 46]. As research continues to advance processing technologies and optimize material properties, cotton stalk-derived packaging is poised to play a key role in the transition toward sustainable materials in the packaging industry.

5.2 *Geotextiles and Agro-Textiles*

Cotton stalks, an abundant lignocellulosic agricultural residue, are increasingly being valorized as raw material for geotextiles and agro-textiles aimed at soil erosion control and the development of biodegradable mulch mats, offering a sustainable alternative to synthetic materials. Recent research demonstrates that natural fiber-based geotextiles, including those derived from cotton stalk bast fibers, exhibit excellent mechanical properties such as tensile strength, elongation, and flexibility, making them suitable for applications in slope stabilization, embankment protection, and water conservation structures1. Modified cotton stalk bast fiber geotextiles, especially when treated with agents like tetraalkoxysilane, show enhanced durability, anti-bursting, and tearing toughness, as well as improved water resistance and barrier performance, which are critical for long-term field deployment in water conservancy and civil engineering projects [47]. Field studies confirm that biological geotextiles constructed from indigenous plant fibers can reduce soil loss by 67–99% and runoff by 26–81% across diverse agro-environments, with their effectiveness attributed to their ability to provide ground cover, increase soil moisture retention, and promote vegetative growth on degraded slopes [48]. The use of cotton stalk-based geotextiles not only stabilizes soil and prevents erosion but also supports the restoration of organic matter and soil structure, contributing to improved crop productivity and land rehabilitation [48, 49]. In agro-textile applications, cotton stalk fibers are processed into nonwoven mulch mats using techniques such as needle punching, resulting in biodegradable products that suppress weed growth, conserve soil moisture, and enhance soil temperature regulation, thereby reducing the need for chemical herbicides and irrigation [50, 51]. Recent advancements highlight the development of mulch mats from end-of-use cotton textiles and cotton stalk fibers, which demonstrate high water absorption, effective weed control, and rapid biodegradation, supporting circularity and nutrient cycling in agricultural systems [50, 51]. The integration of cotton stalks into geotextile and agro-textile products not only addresses the challenge of agricultural waste management but also aligns with global sustainability goals by providing eco-friendly, cost-effective, and high-performance solutions for soil conservation and sustainable agriculture.

5.3 Pulp and Paper Production

Cotton stalks, a widely available agricultural residue, have emerged as a promising non-wood raw material for pulp and paper production, addressing both resource scarcity and environmental concerns associated with traditional wood-based pulping [52, 53]. In mechanical pulping processes, cotton stalks are chipped and subjected to thermo-mechanical treatments, such as high-pressure steaming and mechanical refining, to produce pulp suitable for binderless boards and soft boards, these methods yield products with high mechanical strength, though water absorption can be higher due to the bark content [54, 55]. Mechanical pulping is advantageous for its lower chemical input and energy efficiency, but the resulting fibers may have variable length and purity, often necessitating blending with other pulps like bagasse to achieve desired board properties [54, 55]. Chemical pulping, particularly the kraft and soda processes, is extensively applied to cotton stalks to produce higher-quality pulp for paper and specialty applications [56]. In these methods, cotton stalk chips are digested with alkaline chemicals such as sodium hydroxide for soda pulping and a mixture of sodium hydroxide and sodium sulfide for kraft pulping, under controlled temperature and pressure, effectively removing lignin and hemicellulose to liberate cellulose fibers [56, 57]. Studies report pulp yields of 32–34% for both kraft and soda pulping, with kraft pulping generally providing higher yields and better fiber properties,key operational parameters such as temperature, alkali concentration, and cooking time significantly influence pulp quality and kappa number [56]. The resulting chemical pulps from cotton stalks exhibit good mechanical strength, brightness, and suitability for corrugated boxes, writing, and printing papers, with chemical pulps outperforming semi-chemical pulps in terms of breaking length, burst factor, and tear resistance [52, 56]. Recent advancements include alkaline peroxide extrusion pulping, which combines mechanical and chemical actions to achieve higher yields, improved whiteness, and reduced chemical consumption, further enhancing the environmental profile of cotton stalk pulping [12, 58]. The valorization of cotton stalks in pulp and paper production not only mitigates agricultural waste disposal issues but also supports sustainable industry practices by providing a renewable, cost-effective alternative to wood, with ongoing research optimizing process parameters for maximum efficiency and product quality [52, 53, 59].

6 Environmental and Economic Impact

6.1 Environmental Sustainability

The environmental sustainability of cotton waste is increasingly evaluated through comprehensive life cycle assessment (LCA) methodologies, which quantify the environmental impacts of cotton waste management from production to end-of-life and highlight the critical role of waste reduction and circular economy integration. Recent systematic reviews of LCA studies reveal that recycling and reusing cotton waste significantly reduce greenhouse gas emissions, water consumption, and chemical use compared to the production of virgin cotton fibers, with dynamic LCA models further emphasizing the climate benefits of temporary biogenic carbon storage in cotton-based products [60, 61]. Cotton waste, which constitutes approximately 24% of total textile waste, is generated at every stage of the textile value chain, from spinning and weaving to garment manufacturing, and its effective management is essential for minimizing landfill disposal and environmental pollution [62, 63]. Mechanical and chemical recycling processes enable the transformation of pre- and post-consumer cotton waste into high-value products such as regenerated fibers, composite reinforcements, cellulose nanocrystals, and biofuels, thereby supporting resource efficiency and reducing the ecological footprint of the textile sector [63]. Comparative LCA studies demonstrate that incorporating recycled cotton content, even at low blending ratios, leads to substantial environmental savings across multiple impact categories, including global warming potential, water use, and toxicity, while also diverting waste from landfills and incineration [62, 64]. The integration of cotton waste into circular economy frameworks is further advanced by policy initiatives, such as extended producer responsibility and take-back systems, which incentivize manufacturers to design for recyclability and promote closed-loop material flows [65, 66]. Innovative recycling technologies, including enzymatic and thermal depolymerization, are being developed to enhance the scalability and quality of recycled cotton fibers, while consumer education and industry collaboration are recognized as key drivers for increasing recycling rates and supporting sustainable textile production [64, 66]. Overall, the valorization of cotton waste through LCA-guided strategies and circular economy integration not only mitigates environmental impacts but also fosters economic and social sustainability within the global textile industry.

6.2 Carbon Footprint and Emission Savings

The carbon footprint and emission savings resulting from cotton waste management are significant, with multiple life cycle assessment (LCA) studies confirming that recycling and composting cotton waste can dramatically reduce greenhouse gas (GHG) emissions compared to the production and disposal of virgin cotton products [65], Sustainability [67]. The use of recycled cotton in textile manufacturing avoids the resource- and energy-intensive processes of cotton cultivation, harvesting, and ginning, which are major contributors to emissions in the cotton value chain, wherein fertilizer production alone accounts for nearly half of total GHG emissions in conventional cotton farming [67, 68]. By diverting cotton waste from landfills, where decomposition would otherwise release methane, a potent GHG, recycling and composting strategies further mitigate climate change impacts [67, 69]. Recent LCA research demonstrates that incorporating just 30% recycled cotton into textile products can reduce overall GHG emissions by 2.2–8.6% compared to using 100% virgin cotton, with even greater reductions possible at higher recycled content ratios [65, 67]. Mechanical recycling of cotton waste is substantially less energy-intensive than virgin cotton processing, and the environmental benefits extend to water and land use, as recycled cotton circumvents the need for new agricultural inputs and irrigation [67, 70]. Moreover, composting cotton waste not only avoids landfill emissions but also enables biogenic carbon storage in soils, contributing to a net cooling effect and a 69% reduction in cumulative radiative forcing over a 10-year period compared to landfill or incineration scenarios [71]. Field trials embedding shredded cotton waste into agricultural soils have shown that for every 2.3 metric tons of cotton waste composted, 2.07 metric tons of carbon emissions were avoided, highlighting the potential for soil carbon sequestration and improved nutrient cycling [72]. Although the recycling process itself requires some energy and transportation, the overall life cycle emissions remain far lower than those associated with virgin cotton, and the environmental advantages are clear when considering the full system impacts from production to end-of-life. As cotton recycling technologies advance and circular economy initiatives expand, the carbon footprint of cotton textiles can be substantially reduced, supporting global climate goals and sustainable resource management.

6.3 Economic Benefits

The economic benefits of utilizing cotton waste and cotton stalks as alternative resources in the textile and allied industries are substantial, particularly in terms of reducing dependency on wood pulp and synthetic fibers while simultaneously promoting rural income and waste monetization. By recycling pre- and post-consumer cotton waste into new yarns and fabrics, the textile sector can significantly lower its reliance on virgin cotton and imported short-staple fibers, as demonstrated in recent pilot projects where regenerated cotton yarns replaced a portion of virgin inputs, leading to cost-effective production and averted expenses on fiber imports [73]. This transition not only decreases the demand for wood pulp (traditionally used in man-made cellulosic fibers) and synthetic fibers (derived from petroleum), but also mitigates the environmental and economic risks associated with resource depletion and price volatility in global raw material markets [62, 63]. The valorization of cotton waste through mechanical and chemical recycling processes enables

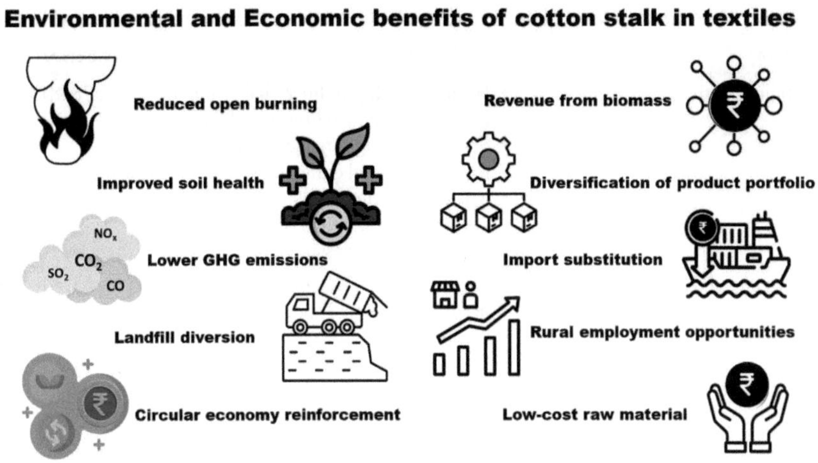

Fig. 1 Environmental and economic benefits of utilizing cotton stalk residue in textile industry. (*Sources* [14, 61, 74–77])

the creation of high-value products such as regenerated yarns, composite reinforcements, and nonwovens, which can directly substitute for virgin materials in a range of applications, further reducing the sector's material costs and environmental footprint [63, 64]. In rural economies, the collection and processing of cotton stalks and textile waste provide new streams of income for farmers and local entrepreneurs, who can supply raw material for biocomposites, packaging, pulp, and other value-added products, thus supporting rural livelihoods and fostering economic diversification [14]. Waste monetization is further enhanced by the development of scalable recycling infrastructure and efficient supply chains, which lower operational costs and increase profitability as recycling processes are optimized and scaled up [73]. Additionally, the conversion of cotton waste into energy products such as bioethanol and briquettes offers substantial savings over traditional fuels, with studies reporting up to 80% reduction in annual fuel costs when cotton-based briquettes replace wood pellets or fuel oil [62]. Overall, integrating cotton waste into circular economy models not only strengthens the economic resilience of the textile industry but also delivers tangible financial benefits to rural communities by transforming agricultural residues and textile discards into valuable resources. Figure 1 enlists various environmental and economic benefits of cotton stalk residue integration in sustainable textile system.

7 Challenges and Limitations

The integration of cotton stalks into textile and allied industries as resource mixes faces a spectrum of challenges and limitations, spanning technical, economic, market, and policy dimensions, which collectively constrain their large-scale adoption and

Table 3 Challenges and limitations of cotton stalks utilization in textile and allied industries

Category	Key challenges	References
Technical challenges	• High variability in fiber quality due to anatomical and environmental factors • Short fiber length reduces suitability for spinning and composites • Residue contamination (soil, agrochemicals, mulch films) complicates processing and quality	[24, 78–81]
Processing and scaling barriers	• High costs for collection, preprocessing, and specialized equipment • Lack of standardized extraction technologies and protocols • Logistical challenges from seasonal availability and dispersed smallholder farms	[11, 55, 81, 82]
Market and consumer acceptance	• Limited consumer awareness and perceived performance issues • Branding and certification limitations • Hesitancy from manufacturers due to concerns about fiber consistency and market positioning	[67, 83]
Policy and regulatory gaps	• Fragmented or misaligned regulatory standards • Lack of targeted policies and infrastructure for by-product valorization • High compliance costs and absence of market-enabling incentives • Regulatory and trade barriers	[55, 67, 84, 85]

sustainable utilization. Table 3 gives a synthesized overview of the major challenges and limitations associated with the integration of cotton stalks into textile and allied industries.

7.1 Technical Challenges

One of the foremost technical barriers is the significant variation in fiber quality derived from cotton stalks, which stems from both intra-plant and inter-plant differences. Research demonstrates that fiber properties such as tensile strength, elongation, diameter, and length differ markedly depending on the anatomical section of the stalk and environmental factors during plant growth, leading to inconsistent material performance in end-use applications [24, 79]. For instance, fibers from the root section often exhibit lower tenacity and maturity compared to those from the middle or top sections, which complicates quality control and standardization in industrial processes [24, 79]. Additionally, short fiber length is a persistent issue, as cotton stalk fibers are generally shorter than conventional cotton lint or bast fibers like jute and flax, limiting their suitability for high-value spinning and textile applications unless advanced extraction and blending methods are employed [81]. This shortness also impacts the mechanical properties of composites and nonwovens, often necessitating the use of binders or reinforcement with longer fibers to achieve desired performance [80]. Residue contamination further complicates processing, as cotton stalks collected from fields are prone to contain soil, dust, and remnants of agrochemicals, as well as potential mixing with mulch films in regions like China, which can hinder fiber separation, reduce yield, and introduce impurities into the final product [24, 78]. These contaminants not only affect the physical and chemical properties of the fibers but also increase the burden on downstream cleaning and refining processes.

7.2 Processing and Scaling Barriers

The processing and scaling of cotton stalk-based fibers are hindered by high initial costs and the absence of standardized, widely adopted technologies. Collection, transportation, and preprocessing of cotton stalks require specialized equipment such as uprooters, shredders, and chippers, which represent substantial capital investments for farmers and processors, particularly in regions where cotton stalk collection is a seasonal activity [55, 82]. Studies indicate that the cost of ready-to-use cotton stalk chips delivered to factories can be significant, with labor, transportation, and equipment expenses collectively raising the price per tonne and impacting the economic feasibility for small and medium enterprises [82]. Furthermore, the lack of uniform processing standards and technology bottlenecks impedes the development of consistent, high-quality fiber products [11]. Existing extraction methods such as mechanical decortication, carding as well as chemical and enzymatic treatments, often yield fibers with variable properties and may not be optimized for industrial-scale throughput or environmental sustainability [11, 81]. The absence of established protocols for detoxification, quality enhancement, and by-product management further limits the efficiency and scalability of cotton stalk valorization [11]. Additionally, logistical challenges in the collection and supply chain, such as the seasonal nature of stalk availability and the need for efficient aggregation from dispersed smallholder farms, add to the complexity and cost of scaling operations [55, 82].

7.3 Market and Consumer Acceptance

The market and consumer acceptance of cotton stalk-based products is constrained by limited awareness, performance perceptions, and branding challenges. Despite the environmental benefits, there is a general lack of consumer familiarity with the properties and advantages of cotton stalk fibers, which affects their willingness to adopt such products [11]. Surveys reveal that consumers strongly associate quality, durability, and comfort with conventional cotton and cotton blends, and may be skeptical of alternative fibers or resource mixes that are perceived as inferior in tactile or aesthetic qualities [83]. Branding limitations are evident, as the market for sustainable or recycled fibers is often fragmented, and the proliferation of eco-labels and claims can create confusion and skepticism among consumers and brands alike [67]. The absence of strong, recognizable brands or certification schemes specifically for cotton stalk-based textiles further hampers market penetration and premium pricing opportunities. Moreover, manufacturers may be reluctant to incorporate cotton stalk fibers into their product lines due to concerns about consistency, performance, and the potential impact on brand reputation, especially in high-end or technical textile segments [11, 83]. These perceptions are compounded by the competitive pressure from established synthetic and bio-based fibers, which often offer superior or more predictable performance characteristics [11].

7.4 Policy and Regulatory Gaps

Policy and regulatory frameworks for the utilization of cotton stalks as resource mixes remain underdeveloped, with notable gaps that hinder widespread adoption. While some national and regional policies promote crop diversification and sustainable resource use, there is often a disconnection between policy intentions and on-the-ground implementation, particularly in cotton-producing countries like India [84]. Regulatory standards for cotton production, harvesting, and residue management are frequently fragmented, with overlapping or inconsistent requirements that create administrative burdens for producers and processors. The complexity and cost of compliance with certification schemes, especially for smallholders, can be prohibitive, reducing incentives for participation and limiting the scale of certified sustainable cotton stalk utilization [67]. Moreover, current policies tend to focus on mono-cropping and cash crop production, overlooking the need for integrated value chains and market development for by-products such as cotton stalk fibers [84]. The lack of targeted incentives, capacity-building programs, and infrastructure support for collection, processing, and quality assurance further constrains the development of a robust supply chain for cotton stalk-based materials [11, 55]. Additionally, regulatory barriers such as quality control orders and non-tariff barriers can restrict access to raw materials and specialized fiber varieties, impeding innovation and competitiveness in the sector [85].

8 Conclusion

The investigation into the utilization of cotton stalk residue for textile resource mixes reveals a compelling opportunity to realign textile manufacturing with sustainability goals. Key findings underscore the high cellulose content, mechanical robustness, and compatibility of cotton stalk fibers with existing fiber blends. Advanced processing techniques (mechanical, chemical, and biotechnological) enhance fiber extraction efficiency and quality, making cotton stalks suitable for applications ranging from nonwoven insulation to blended yarns and biodegradable packaging. Their integration not only reduces agricultural waste but also contributes to emission reductions, resource circularity, and rural economic development. Reinforcing cotton stalks as a sustainable solution, the chapter presents them as an ecologically sound and economically feasible alternative to wood-derived and petroleum-based fibers. Their use promotes biodiversity preservation, reduces dependence on land- and water-intensive crops, and supports the circular economy by converting agricultural residues into high-value materials. Applications in geotextiles, composites, and pulp further diversify their potential, while life cycle assessments confirm lower environmental footprints compared to conventional materials. Looking ahead, cotton stalks hold transformative potential in shaping the future of eco-textiles. However, several avenues warrant attention for successful integration:

(i) Research Needs: Greater investment is required in material science and process optimization to standardize fiber quality, improve pretreatment methods, and ensure scalability. Advanced biotechnology applications can enhance fiber yield and fineness.

(ii) Policy Incentives and Industrial Collaborations: Government support in the form of subsidies, tax incentives, and inclusion of cotton stalk utilization in agricultural waste management policies can accelerate adoption. Collaborative models between textile firms, research institutions, and farming cooperatives are essential for streamlined supply chains.

(iii) Market Development and Awareness: Consumer education and marketing strategies should emphasize the environmental and performance benefits of cotton stalk-based products. Eco-labels and certifications specific to such fibers can foster consumer trust and brand differentiation.

(iv) Integration into Sustainability Frameworks: Cotton stalk fibers must be embedded within broader sustainability frameworks such as circular design, cradle-to-cradle manufacturing, and extended producer responsibility. Their inclusion can make textile production more resilient, ethical, and eco-conscious.

In conclusion, valorizing cotton stalks is not merely a waste management strategy but a forward-looking approach that aligns with global sustainability transitions. Their thoughtful integration into textile value chains can redefine the material base of the industry, thus making sustainability not an option, but a foundation.

References

1. Abbate S, Centobelli P, Cerchione R, Nadeem SP, Riccio E (2024) Sustainability trends and gaps in the textile, apparel and fashion industries. Environ Dev Sustain 26(2):2837–2864
2. Leal Filho W, Perry P, Heim H, Dinis MAP, Moda H, Ebhuoma E, Paço A (2022) An overview of the contribution of the textiles sector to climate change. Front Environ Sci 10:973102
3. Häuser L, Ujawane P (2025) Under pressure: the textiles & apparel industry's multiple challenges. ISS ESG. https://insights.issgovernance.com/posts/under-pressure-the-textiles-apparel-industrys-multiple-challenges/. Accessed July 2025
4. Tian W, Huang K, Zhu C, Sun Z, Shao L, Hu M, Feng X (2022) Recent progress in biobased synthetic textile fibers. Front Mater 9:1098590
5. Islam T, Hossain MM, Covington SM (2025) Natural fibers towards fashion sustainability: a review of raw materials, production, application, and perspective. J Nat Fibers 22(1):2462218
6. Frazier RM, Lendewig M, Vera RE, Vivas KA, Forfora N, Azuaje I, Gonzalez R (2024) Textiles from non-wood feedstocks: challenges and opportunities of current and emerging fiber spinning technologies. J Bioresour Bioprod
7. Choudhury K, Tsianou M, Alexandridis P (2024) Recycling of blended fabrics for a circular economy of textiles: separation of cotton, polyester, and elastane fibers. Sustainability 16(14):6206
8. Duhoux T, Le Blévennec K, Manshoven S, Grossi F, Arnold M, Mortensen LF (2022) Textiles and the environment: the role of design in Europe's circular economy (ETC/CE Report 2/2022). European Environment Agency
9. Wardiningsih W, Syahputra IP, Rudy R, Mulyani RWE, Pradana SM, Pudjiati P, Sumihartati A (2024) Exploring eco-friendly pseudo stem fiber from the agricultural byproducts of the Alpinia purpurata plant. Bioresour Technol Rep 27:101947
10. Karimah A, Ridho MR, Munawar SS, Adi DS, Damayanti R, Subiyanto B, Fudholi A (2021) A review on natural fibers for development of eco-friendly bio-composite: characteristics, and utilizations. J Market Res 13:2442–2458
11. Zhang Z, Wei X, Gao L, Wang Z (2025) Multi-dimensional utilization of cotton and its residues: current status and future directions. J Text Sci Fashion Technol 11(4):Article 000769. https://doi.org/10.33552/JTSFT.2025.11.000769
12. Prakash S, Sharma K, Dhumal S, Senapathy M, Deshmukh VP, Kumar S, Kumar M (2024) Unlocking the potential of cotton stalk as a renewable source of cellulose: a review on advancements and emerging applications. Int J Biol Macromol 261:129456
13. Dong Z, Hou X, Sun F, Zhang L, Yang Y (2014) Textile grade long natural cellulose fibers from bark of cotton stalks using steam explosion as a pretreatment. Cellulose 21:3851–3860
14. Ibrahim RA, Inan H, Fahim IS (2023) A comparative cradle-to-gate life cycle assessment of three cotton stalk waste sustainable applications. Sci Rep 13(1):20781
15. MacDonald S, Lanclos K, Meyer L, Soley G (2024) The world and United States cotton outlook. Presented at the 100th Annual Agricultural Outlook Forum, U.S. Department of Agriculture. https://www.usda.gov/sites/default/files/documents/2024AOF-cotton-outlook.pdf
16. Meyer L, Dew T (2023) Cotton and wool outlook: December 2023 (Situation and Outlook Report CWS-23k). U.S. Department of Agriculture, Economic Research Service. https://www.ers.usda.gov/sites/default/files/_laserfiche/outlooks/108086/CWS-23k.pdf
17. Singh A (ed) (2024) URJA project: transforming cotton stalk waste into bioenergy, biochar, and carbon credits. Cotton Statistics and News, Cotton Association of India, (no 25)
18. Belem A, Ouédraogo-Koné S, Lankoandé YF, Koulibaly B, Sakandém F, Bamogo A, Troré M (2024) Evaluation of the forage potential of harvest residues in cotton farms in Western Burkina-Faso. Afr J Food Agric Nutr Dev 24(6):26533–26553
19. Parmar PR, Rajkumar BK, Malek N, Ukani H (2025) Comprehensive elucidation of structural and compositional changes in cotton stalks undergoing biodegradation by potent cellulolytic bacteria. Cellul Chem Technol 59(1–2):111–119
20. Zhong P, Chen P, Huo P, Ma L, Xu Z, Li F, Cai C (2025) Characterization of cotton stalk as a lignocellulosic feedstock for single-cell protein production. Biores Technol 417:131797

21. Jindal M, Kumar A, Rawat S, Thallada B (2023) Lignin-first biorefinery approach for the valorization of cotton stalks to phenolic monomers. Sustain Energy Fuels 7(16):3926–3938
22. Phadtare P, Kalbande S (2024) Thermal behaviour, chemical composition and morphological analysis of cotton stalk for efficient biochar production. J Agric Eng (India) 61(4):1–12
23. Xu F, Shang J, Abdurexit A, Jamal R, Abdiryim T, Li Z, Huang L (2024) Effect of chemical treatment of cotton stalk fibers on the mechanical and thermal properties of PLA/PP blended composites. Polymers 16(12):1641
24. Nkomo NZ, Nkiwane LC (2017) Statistical analysis of the significance of variation in properties of cotton stalk fibres extracted from different sections of the cotton stalk. IOSR J Eng (IOSRJEN) 7(4):53–62
25. Zhao W, Xie J, Wang Z, Gao Q, Chen M (2022) Investigation of mechanical properties of cotton stalk based on multi-component analyses. Int Agrophysics 36(4):257–267
26. Dong Z, Li N, Chu T, Ding J, Zhang J, Dong A (2022) High-quality natural fibers from cotton stalk bark via limited alkali penetration and simultaneous accelerated temperature rise. Materials 15(2):422
27. Reddy N, Yang Y (2009) Properties and potential applications of natural cellulose fibers from the bark of cotton stalks. Biores Technol 100(14):3563–3569
28. Elfaleh I, Abbassi F, Habibi M, Ahmad F, Guedri M, Nasri M, Garnier C (2023) A comprehensive review of natural fibers and their composites: an eco-friendly alternative to conventional materials. Results Eng 19:101271
29. Farag GY, Abou Taleb E, Hamouda T (2024) Natural fibers extraction methods and properties: a review. Egypt J Chem 68(3):445–464
30. Hou X, Sun F, Zhang L, Luo J, Lu D, Yang Y (2014) Chemical-free extraction of cotton stalk bark fibers by steam flash explosion. BioResources 9(4):6950–6967. https://digitalcommons.unl.edu/textiles_facpub/42
31. Li YY, Du GM, Feng XJ, Mu YW, Li DQ (2016) Extraction and hydrophobic modification of cotton stalk bark fiber. Int J Polym Sci 2016(1):8769360
32. Dong Z, Hou X, Haigler I, Yang Y (2016) Preparation and properties of cotton stalk bark fibers and their cotton blended yarns and fabrics. J Clean Prod 139:267–276
33. Jawahar V, Gabriel M, Santhanam S, Selvaraj SK (2023) Sustainable waste cotton and pigeon pea stalk fibers composite materials for acoustics and thermal properties. J Eng Fibers Fabr 18:15589250231189814
34. Mohammadi M, Taban E, Tan WH, Din NBC, Putra A, Berardi U (2024) Recent progress in natural fiber reinforced composite as sound absorber material. J Build Eng 84:108514
35. Subramanian A, Selvaraj SK, Mani R, Chandrasekaran P, Kunjuraman S, Guru R, Santhanam S (2024) Utilizing waste cotton/pigeon pea stalk fibers composites for enhanced sound absorption and insulation in automotive interiors. J Nat Fibers 21(1):2333940
36. Jiang N (2008) Fabrication and evaluation of specialty nonwoven materials. Doctoral dissertation, Louisiana State University and Agricultural & Mechanical College. LSU Digital Commons. https://doi.org/10.31390/gradschool_dissertations.723
37. Jiang N, Chen JY, Parikh DV (2009) Acoustical evaluation of carbonized and activated cotton nonwovens. Biores Technol 100(24):6533–6536
38. Zhang D, Zhou X, Gao Y, Lyu L (2022) Structural characteristics and sound absorption properties of waste hemp fiber. Coatings 12(12):1907
39. Ahmad W, McCormack SJ, Byrne A (2025) Biocomposites for sustainable construction: a review of material properties, applications, research gaps, and contribution to circular economy. J Build Eng 105:112525
40. Chakraborty M (2025) Automotive applications of biocomposites–a review. Int Res J Mod Eng Technol Sci 7(1):2636–2645. https://doi.org/10.56726/IRJMETS66376
41. Ullah S, Akhter Z, Palevicius A, Janusas G (2025) Natural fiber-based biocomposites for potential advanced automotive applications. J Eng Fibers Fabr 20:15589250241311468
42. Jabu MA, Alugongo AA, Nkomo NZ (2024) Application of natural fibre composites in interior panels in the automotive industry: a review. Int J Eng Trends Technol 72:91–98

43. Kamath MG (2004) Processing and evaluation of cotton-based composites for automotive and other applications. Master's thesis, University of Tennessee. TRACE: Tennessee Research and Creative Exchange. https://trace.tennessee.edu/utk_gradthes/2261
44. Song Z, Lian X, Ran J, Zheng X, Wang X, Lian X (2025) Effect of cotton stalk biochar content on the properties of cotton stalk and residual film composites. Agriculture 15(12):1243
45. Paswan M, Adhikary S, Salama HH, Rusu AV, Zuorro A, Dholakiya BZ, Bhattacharya S (2023) Microbial synthesis of lactic acid from cotton stalk for polylactic acid production. Microorganisms 11(8):1931
46. Cai C, Wang Z, Ma L, Xu Z, Yu J, Li F (2024) Cotton stalk valorization towards bio-based materials, chemicals, and biofuels: a review. Renew Sustain Energy Rev 202:114651
47. Xu H (2019) Modified cotton stalk bast fiber geotextiles of a kind of tetraalkoxysilane and preparation method thereof (Patent no.: CN109868555A). Nantong Textile Vocational Technology College, China. https://patents.google.com/patent/CN109868555A/en
48. Bhattacharyya R, Fullen MA, Booth CA, Kertesz A, Toth A, Szalai Z, Toan TT (2011) Effectiveness of biological geotextiles for soil and water conservation in different agro-environments. Land Degrad Dev 22(5):495–504
49. Pal A, Adhikary R, Bera M, Garanayak R, Gaikwad DJ, De SK (2020) An overview of geotextiles in agriculture. Biosci Biotechnol Res Commun 13(12):15–18. https://bbrc.in/wp-content/uploads/2021/02/Special-Issue-Proof-003.pdf
50. Cao H, Ludwig K, Cobb K, Scott C, Jin Y, Knight B, Shen H (2024) Development and evaluation of biodegradable weed control mulch mats from end-of-use cotton waste. Sustainability 16(16):7083
51. Cao H, Ludwig K, Cobb K, Scott C, Jin Y, Knight B, Burrichter M, Shen H (2024a) Development and research of sustainable mulch mats from end-of-use cotton textiles. In: International textile and apparel association annual conference proceedings, vol 80, no 1. https://doi.org/10.31274/itaa.17158
52. Pandey SN, Shaikh AJ (1987) Utilisation of cotton plant stalk for production of pulp and paper. Biol Wastes 21(1):63–70
53. Worku LA, Bachheti A, Bachheti RK, Rodrigues Reis CE, Chandel AK (2023) Agricultural residues as raw materials for pulp and paper production: overview and applications on membrane fabrication. Membranes 13(2):228
54. Li LX, Chen HT (2014) Study on the physical and chemical and mechanical pulping characteristic of cotton stalk fiber. Adv Mater Res 884:158–161
55. Saxena S (2024) Cotton plant stalks: a valuable industrial raw material. *Cotton Statistics & News, 2024–25*(11), 1–3. Cotton Association of India. https://caionline.in
56. Erana LA (2021) Pulp production from cotton stalks. J Text Eng Fash Tech 3(1):23–28
57. Akgül M, Tozluoglu A (2009) A comparison of soda and soda-AQ pulps from cotton stalks. Afr J Biotech 8(22):6127–6133
58. Hu Z, Nie X (2014) Alkaline peroxide extrusion pulping of cotton bast and cotton stalk. BioResources 9(2):2856–2865
59. La Rubia MD, Jurado-Contreras S, Navas-Martos FJ, García-Ruiz Á, Morillas-Gutiérrez F, Moya AJ, Rodríguez-Liébana JA (2024) Characterization of cellulosic pulps isolated from two widespread agricultural wastes: cotton and sunflower stalks. Polymers 16(11):1594
60. Pires ST, Williams A, Daystar J, Sagues WJ, Lan K, Venditti RA (2024) Evaluating cotton apparel with dynamic life cycle assessment: the climate benefits of temporary biogenic carbon storage. BioResources 19(3):5074–5095
61. Vitale GS, Iacuzzi N, Zingale S, Lombardo S, Tuttolomondo T, Guarnaccia P (2025) Environmental sustainability of cotton: a systematic literature review of life cycle assessments. J Agric Food Res 22:102069
62. Johnson S, Echeverria D, Venditti R, Jameel H, Yao Y (2020) Supply chain of waste cotton recycling and reuse: a review. AATCC J Res 7(1S):19–31
63. Lu L, Fan W, Meng X, Xue L, Ge S, Wang C, Lam SS (2023) Current recycling strategies and high-value utilization of waste cotton. Sci Total Environ 856:158798

64. Khan IA, Malik MH, Fraz A, Ibrahim H (2025) From waste to weave: recycling industrial and consumer cotton waste for sustainable and cost-effective terry fabric production. Tekstilec 68(1):42–56
65. Roy S, Chu YYJ, Chopra SS (2023) Life cycle environmental impact assessment of cotton recycling and the benefits of a Take-Back system. Resour Conserv Recycl Adv 19:200177
66. Sabareesh N, Karthik S, Nagarajan L (2024) Strategies for reducing, reusing and recycling textiles within a circular economy framework. Int J Eng Res Technol (IJERT) 13(11). https://doi.org/10.17577/IJERTV13IS110097
67. Sustainability Directory (2025) What is recycled cotton's carbon footprint? Fashion Sustainability Directory. https://fashion.sustainability-directory.com/question/what-is-recycled-cottons-carbon-footprint/. Accessed 5 July 2025
68. Vandepaer L (2024) The carbon footprint of cotton. Carbonfact. https://www.carbonfact.com/blog/knowledge/carbon-footprint-cotton. Accessed 5 July 2025
69. Pires S (2023) Cotton to compost: transform waste management in the textile industry. Cotton Today. Cotton Incorporated. https://cottontoday.cottoninc.com/cotton-to-compost-transform-waste-management-textile-industry/. Accessed 5 July 2025
70. ZigZagZurich (2024) Recycled cotton–how it is made and what are the benefits for home textiles. ZigZagZurich. https://zigzagzurich.com/blogs/news/recycled-cotton-how-it-is-made-and-what-are-the-benefits-for-home-textiles. Accessed 5 July 2025
71. Pires S (2024) The untapped climate benefit of cotton: exploring its carbon storage potential. Cotton Today. Cotton Incorporated. https://cottontoday.cottoninc.com/climate-benefit-of-cotton-carbon-storage/. Accessed 5 July 2025
72. Pau K (2023) Recycling cotton waste by putting it back in the soil. Cool Hunting. https://coolhunting.com/culture/recycling-cotton-waste-by-putting-it-back-in-the-soil/. Accessed 5 July 2025
73. Von Daniels C, Papú Carrone N, Steenmeijer M, Verstraeten-Jochemsen J (2020) RE.ACT: environmental and economic assessment of post-industrial cotton waste recycling (C. F. Blanco, Review). Circle Economy & UNIDO. https://pro.refashion.fr/eco-design/sites/default/files/fichiers/RE.ACT%20%20Environmental%20and%20Economic%20assessment%20of%20post-industrial%20cotton%20waste%20recycling.pdf
74. Arude VG, Shukla SK, Patil PG, Deshmukh PS (2019) Sustainable development in cotton sector through industrial applications of value added products from cotton stalk. Cotton Res J 10(1):1
75. Makavana JM, Chauhan PM, Lakhani AL, Balas PR, Sarsavadia PN (2025) Characterization of physical and chemical properties of cotton stalk and its derived bio-char. Plant Arch 25:433–442. https://doi.org/10.51470/PLANTARCHIVES.2025.SP.ICTPAIRS-063
76. Raju NP, Raju M (2019) Experimental study of cotton stalk pellet renewable energy potential from agricultural residue woody biomass as an alternate fuel for fossil fuels to internal combustion engines. Int J Eng Res Technol 8:746–750
77. Rizal S, HPS AK, Oyekanmi AA, Gideon ON, Abdullah CK, Yahya EB, Rahman AA (2021) Cotton wastes functionalized biomaterials from micro to nano: a cleaner approach for a sustainable environmental application. Polymers 13(7):1006
78. Baikeli Y, Xu H, Zhang J, Yue Y, Guo G (2025) Innovations and challenges in mechanized cotton stalk harvesting for environmental sustainability in the field: a review. J ASABE 68(3):365–377
79. Kothari N, N (2012) Improvement of cotton fiber maturity and assessment of intra-plant fiber variability. Doctoral dissertation, Texas A&M University.. Texas A&M University. https://oaktrust.library.tamu.edu/server/api/core/bitstreams/cc9ee0a9-90e1-480a-a6fb-482c1b7d1358/content
80. Li G, Yu Y, Zhao Z, Li J, Li C (2003) Properties study of cotton stalk fiber/gypsum composite. Cem Concr Res 33(1):43–46
81. Liu L, Jin S (2009) U.S. Patent No. 7,624,478. United States Patent and Trademark Office
82. Shaikh AJ, Gurjar RM, Patil PG, Paralikar KM, Varadarajan PV, Balasubramanya RH (2017) Particle boards from cotton stalk. Central Institute for Research on Cotton Technology (CIRCOT). https://staging.icac.org/tis/regional_networks/asian_network/meeting_5/documents/papers/PapShaikhA.pdf

83. Thompson E (2012) Diluting cotton content compromises quality and brand perception. Consumer Facts from Cotton Incorporated Lifestyle Monitor™, as featured in *The Robin Report*. https://therobinreport.com/diluting-cotton-content-compromises-quality-and-brand-perception/. Accessed 7 July 2025
84. Keller C, Joshi S, Joshi T, Goldmann E, Riar A (2024) Challenges for crop diversification in cotton-based farming systems in India: a comprehensive gap analysis between practices and policies. Front Agron 6:1370878
85. UNCTAD (2025) Trends in natural and man-made fibres trade. United nations conference on trade and development. https://unctad.org/publication/trends-natural-and-man-made-fibres-trade
86. Tatsiopoulos IP, Tolis AJ (2003) Economic aspects of the cotton-stalk biomass logistics and comparison of supply chain methods. Biomass Bioenerg 24(3):199–214

The Future of Textile Resource Mix: Towards a Regenerative Industry

Swikruti Pradhan and Asimananda Khandual

Abstract The textile industry is at a critical crossroad, where sustainability alone is no longer sufficient; a shift toward the regeneration of resources is essential. The major stakeholders in the textile sector need to examine the limitations of conventional textile production, including environmental degradation, over-reliance on finite resources, and challenges in waste management, and highlight emerging regenerative approaches that can redefine the industry. Innovative resource mixes in textiles can drive this transformation, which is the need of the hour. The future of textiles lies in regeneration and resource mix, moving beyond sustainability to actively restore ecosystems, enhance biodiversity, and minimise waste. This chapter, *The Future of Textile Resource Mix: Towards a Regenerative Industry*, examines the limitations of conventional fibre choices such as natural, synthetic, and blended, while highlighting regenerative alternatives that promote ecological balance. The chapter envisions a regenerative textile industry that restores rather than depletes, urging collaboration between fashion, agriculture, and biotechnology to create a net-positive impact.

Keywords Resource mix · Regenerative textiles · Regenerative industry · Circular economy · Circular materials · Bio-engineering · Alternative fibres

1 Introduction

The global textile and apparel industry stands at the limen of transformative change. As one of the most resource-intensive sectors, its dependency on finite raw materials, unsustainable production practices, and linear supply chains has contributed significantly to environmental degradation, climate change, and socio-economic inequalities. The current textile *resource mix* comprising a blend of natural fibres such as

S. Pradhan (✉)
Rustic Hue®, Bangalore, India
e-mail: swikruti@rustichue.com

A. Khandual
Odisha University of Technology and Research (Formerly CET), Bhubaneswar, Odisha, India
e-mail: asimte@outr.ac.in

S. S. Muthu (ed.), *Resource Mixes in Textiles*, Sustainable Textiles: Production, Processing, Manufacturing & Chemistry, https://doi.org/10.1007/978-3-032-10921-7_4

cotton, wool, silk etc.; synthetic fibres such as polyester and nylon, and regenerated or semi-synthetic fibres including viscose and lyocell is deeply embedded in an extractive model that prioritises economic output over ecological balance [1]. Amid growing climate crises and resource scarcity, the future of textile resources must shift from sustainability as mere mitigation to regeneration as an ethos of healing, restoration, and circularity.

A *resource mix* in textiles refers to the combination of raw materials, chemicals, water, and energy inputs that go into the creation of fibres, yarns, fabrics, and finished garments. This mix directly influences not only the environmental footprint of textile products but also their potential for recyclability, biodegradability, and alignment with circular systems [2]. Historically, the industrial revolution catalysed a gradual shift from handwoven, locally sourced textiles to mass-produced, globally traded materials, increasingly dependent on fossil fuels and synthetic chemistry. While the sustainable fabric segment is still growing with an expectation of reaching USD 161.6 billion by 2037 from USD 36 billion in 2025, with a CAGR of 13% (Elevate Textiles 2025), approximately 63% of global fibre production is synthetic, dominated by polyester, a petroleum-derived material [3]. While synthetic fibres offer affordability and performance benefits, they come with severe ecological costs, plastic pollution, microfibers in oceans, and dependency on non-renewable resources.

Natural fibres, although biodegradable, have their drawbacks. Conventional cotton, for instance, accounts for 2.5% of global agricultural land but uses 16% of the world's insecticides [4]. Wool production, unless managed regeneratively, contributes to methane emissions and land degradation. Therefore, a *regenerative* approach to textile resource mix does not imply merely replacing synthetic fibres with natural ones but rather rethinking how materials are grown, harvested, processed, and disposed of, with an emphasis on ecological restoration, carbon sequestration, and social equity.

Emerging concepts of *regenerative agriculture* involving practices such as cover cropping, no-till farming, holistic grazing, and agroforestry are gaining momentum in fibre cultivation. These methods rebuild soil organic matter and restore degraded soil biodiversity, which can, in turn, enhance carbon drawdown and resilience to climate impacts [5]. Regenerative cotton and wool are already being piloted in parts of India, Australia, and the United States, showing promise as viable alternatives to conventional farming models [6]. When integrated with local economies and indigenous knowledge systems, such practices can also support cultural continuity and rural livelihoods.

On the innovation frontier, *bioengineered materials* such as mycelium leather, bacterial cellulose, and lab-grown silk represent a revolutionary shift in resource creation—where materials are grown rather than extracted. These alternatives offer new functionalities, lower environmental footprints, and the potential to decouple material production from land and water use [7]. Additionally, materials derived from agricultural and food waste e.g. pineapple fibre (Piñatex), orange fibre, and coconut coir, exemplify closed-loop systems that convert waste into value-added products, embodying the principles of circularity and regeneration.

Transitioning to a regenerative textile economy requires more than just material substitution. It demands a systemic transformation of how the industry designs, produces, distributes, and consumes. A regenerative industry is not merely sustainable but *restorative by design*, where each stage of the textile lifecycle contributes to ecosystem health, climate resilience, and community empowerment. It also entails a redefinition of value, shifting from throughput and profit maximisation to regeneration, longevity, and socio-ecological harmony [8].

Policymakers, manufacturers, suppliers, brands, and consumers all play a role in enabling this transition. Regulatory frameworks like the EU Strategy for Sustainable and Circular Textiles (2022) and India's National Technical Textiles Mission (2020) signal institutional support for innovation and sustainability. Brands are increasingly being held accountable through *Extended Producer Responsibility (EPR)* laws, supply chain transparency regulations, and ESG reporting. Consumers, in turn, are becoming more conscious of their consumption patterns, demanding products that are ethically sourced, environmentally friendly, and circular in nature.

The regenerative textile movement also recognises the importance of *interdisciplinary collaboration* between designers, scientists, farmers, technologists, and indigenous communities to co-create solutions that are not only technically sound but culturally rooted. Organisations such as the Sustainable Apparel Coalition, Textile Exchange, Canopy, and Fibershed are at the forefront of preaching for regenerative materials and systems thinking in the textile industry.

This chapter provides a broad analysis of the evolving dynamics of textile resource use and outlines a roadmap for the industry to transition towards a regenerative future. Through an exploration of regenerative natural fibres, cutting-edge bio-based materials, circular resource flows, and enabling mechanisms such as traceability, innovation ecosystems, and policy support, it outlines how the textile industry can evolve from a linear and extractive model to a *net-positive, regenerative system*. Thus, it invites stakeholders across the textile value chain to reimagine resource stewardship as a regenerative practice that sustains and actively heals our planet.

2 Rethinking Textile Resources

The evolution of textile resources has closely mirrored the trajectory of industrial development, technological advancement, and consumer culture. Historically, communities worldwide used locally available materials, e.g. cotton in India, flax in Egypt, wool in Europe, and bark or bast fibres in parts of Africa and Asia, crafting textile systems embedded in ecological rhythms and social rituals. However, the contemporary textile economy has largely dissociated material sourcing from ecological context, leading to extractive practices and a reliance on non-renewable inputs. As the industry grapples with intersecting crises such as climate change, biodiversity loss, water scarcity, and social inequities, it becomes imperative to rethink textile resources through the lens of regeneration.

2.1 *Moving from Linear to Regenerative Thinking*

Traditional industrial textile systems operate linearly: extract, produce, consume, and dispose. This model depletes natural resources and generates massive waste estimated at over 92 million tonnes annually [1]. Moreover, 87% of total fibre input used for clothing is landfilled or incinerated [1]. Regenerative thinking challenges this paradigm by viewing resources as part of a living system that must be nurtured and restored. In a regenerative model, materials are cultivated, processed, used, and returned in ways that enhance ecosystem health, biodiversity, and community well-being [9].

Regeneration aims to create *net-positive outcomes.* It emphasises soil health, water cycles, carbon drawdown, and holistic supply chain relationships. For textiles, this involves reassessing the origin, production, and afterlife of every fibre and chemical used, not only through lifecycle assessment but through a systemic perspective on social and ecological interdependencies (Fletcher 2016).

2.2 *Diversifying and Localising the Fibre Basket*

The current global fibre market is dominated by a handful of resources: polyester (over 50%), cotton (nearly 25%), and regenerated cellulose fibres such as viscose [3]. This heavy concentration poses ecological and geopolitical risks. Polyester's fossil fuel reliance links it to climate change and plastic pollution, while cotton's water use and chemical inputs have degraded vast agricultural landscapes and over-reliance on monocultures, both biological and industrial, erodes resilience.

Diversifying the textile fibre basket with underutilised, *regionally appropriate resources* is a crucial strategy for regeneration. Indigenous and local fibres such as nettle (Girardinia diversifolia) in the Himalayas, banana fibre in Southeast Asia, agave in Mexico, and yak wool in Mongolia provide biodegradable, renewable alternatives rooted in cultural heritage and agroecological knowledge [10]. These fibres often grow without synthetic inputs and require less processing, making them more compatible with circular systems. Localisation also reduces the carbon footprint of transportation and reconnects production with place-based stewardship. Initiatives like *Fibershed's Regional Fibre Systems* in the United States demonstrate how decentralised, bioregional textile economies can support regenerative agriculture, local artisanship, and ecosystem services [6].

2.3 Regenerative Natural Fibres

Natural fibres are generally seen as inherently sustainable, but their ecological value depends completely on how they are cultivated and processed. Regenerative agriculture has emerged as a vital method for restoring degraded lands, sequestering carbon, and producing resilient fibre crops. Regenerative cotton farming, for instance, includes intercropping, compost application, cover cropping, and minimal tillage to increase soil organic matter and reduce dependence on chemical inputs [5].

Wool and other animal fibres offer similar opportunities. Through managed grazing and rotational pasture systems, sheep and alpacas can help regenerate grasslands, enhance soil microbial life, and improve water infiltration [11]. Such practices are especially relevant in semi-arid zones, where traditional herding systems align well with regenerative principles.

The scalability of regenerative natural fibres depends on policy support, supply chain redesign, and capacity building among farmers. Programs like the *Regenerative Organic Certified (ROC)* and *Textile Exchange's Regenerative Outcomes Framework* are paving the way for standardisation and traceability.

2.4 Next-Gen Bio-Based Materials

While regenerative agriculture works in harmony with nature, *biodesign and biofabrication* introduce new paradigms for material production by harnessing the programming capabilities of biology itself. Next-gen bio-based materials such as *mycelium (fungus-based leather), bacterial cellulose,* and *lab-grown silk proteins* offer radically different resource inputs. These materials can be cultivated in controlled environments with minimal use of land, water, or chemicals [7].

Mycelium, for instance, grows on agricultural waste and can be moulded into durable, compostable leather-like sheets. Companies like MycoWorks and Ecovative are scaling this technology as a sustainable alternative to plastic and animal leather. Similarly, Spiber (Japan) and Bolt Threads (USA) are developing protein-based textiles that mimic the properties of silk and wool using yeast fermentation. These technologies raise new questions about scalability, lifecycle impacts, and cost. However, their greatest potential lies in *decoupling material production from extractive systems* and designing fibres that fit into regenerative metabolisms, either through biodegradability or biological recyclability.

2.5 Agro-Waste and Industrial By-Products

Another critical domain for rethinking resources is the use of *agricultural and industrial waste* as textile feedstock. Agro-waste fibres convert what was once considered residue into high-value materials. Examples include: Piñatex, a leather alternative made from pineapple leaf waste; Orange Fibre, derived from citrus juice industry by-products; and Cocona®, made from coconut husk waste.

These materials reduce waste and create new income streams for farmers and food processors. More importantly, their production often avoids the use of virgin land, pesticide application, or intensive water use, making it synergistic with regenerative goals. Textile waste: pre-consumer and post-consumer should also be reconsidered as a *valuable resource.* Innovations in chemical recycling (e.g., Infinited Fibre, Renewcell) now allow for the regeneration of cotton-rich waste into virgin-quality fibres. These technologies, when paired with regenerative sourcing, can create closed-loop systems where waste becomes the starting point for new materials.

2.6 Cultural Regeneration and Material Relationships

Rethinking textile resources is not only a technical or ecological endeavour, but also a cultural and philosophical one. Indigenous and craft communities often hold deep knowledge of regenerative resource use, rooted in seasonal cycles, reciprocity, and multispecies kinship [12]. From hand-spun khadi in India to backstrap weaving in Latin America, these practices offer models of sufficiency, care, and continuity.

The regenerative textile movement challenges the dominant logics of fast fashion, extractivism, and consumerism by re-centring cultural narratives. It invites us to view materials not as commodities but as carriers of relationships between *humans and nature, past and future, labour and landscape.* For systemic change, incorporating these values into design education, sourcing strategies, and brand storytelling is essential.

2.7 Towards Resource Resilience

Rethinking textile resources means diversifying material inputs, embracing regenerative agriculture, harnessing bio-based innovations, valorising waste, and re-rooting production in ecological and cultural contexts. It demands a shift from linear, globalised supply chains to regenerative, decentralised, and circular systems. Such a transition is complex but necessary. It involves policy reforms, consumer education, design innovation, and financial investment in regenerative infrastructure. The future of textile resources will depend not only on technological breakthroughs but on a fundamental reorientation of values towards care, interconnection, and regeneration (Fig. 1).

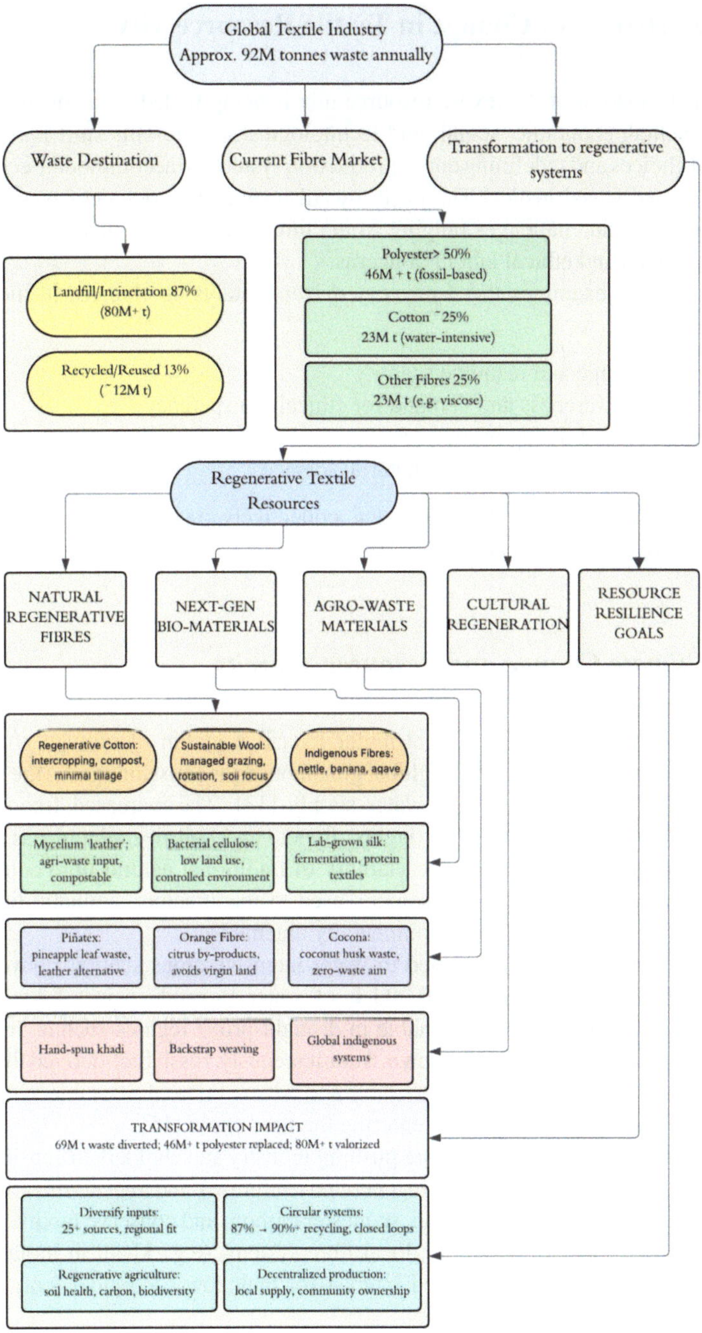

Fig. 1 Bird's eye view on rethinking textile resources

3 Key Drivers of Change in Textile Resource Mix

The transformation of the textile resource mix is being fueled by a convergence of environmental, economic, social, and technological forces. This shift is reshaping material choices and redefining entire production systems to accommodate ecological resilience and social justice. The push toward regenerative practices is a response to complex global challenges ranging from climate change and biodiversity loss to waste overload and ethical labour concerns.

This section examines the *4 primary drivers* catalysing changes in the textile resource mix:

1. Climate change and resource scarcity
2. Consumer awareness and demand for ethical transparency
3. Regulatory and policy frameworks
4. Technological innovation and material science.

Each of these drivers is interconnected, collectively creating momentum towards a regenerative textile economy.

3.1 Climate Change and Resource Scarcity

The textile industry is both a contributor to and a victim of climate change. It contributes approximately 10% of global greenhouse gas emissions and is the second-largest consumer of the world's water supply [13]. The increased frequency of climate-related events, such as droughts, floods, and soil degradation, as well as biodiversity collapse, threatens the viability of traditional natural fibre cultivation. At the same time, finite fossil resources used to manufacture synthetic fibres are becoming economically and environmentally untenable.

Textile production is deeply tied to water-intensive crops such as conventional cotton, which requires 10,000–20,000 L of water to produce just 1 kg of fibre [14]. This is particularly unsustainable in drought-prone regions such as India and Pakistan, where cotton is widely grown. Simultaneously, fossil fuel-derived fibres like polyester and nylon are non-biodegradable and release microplastics into waterways, impacting marine ecosystems and entering the food chain [15].

In response, climate pressures are pushing industry stakeholders to consider *low-impact fibres* (e.g., hemp, linen, regenerated cellulose) and *regenerative agricultural systems* that restore soil health, sequester carbon, and conserve biodiversity. In regions with climate vulnerability, local fibre systems (e.g., khadi in India, alpaca in the Andes) are gaining renewed interest for their low ecological footprint and community-driven models.

3.2 *Consumer Awareness and Demand for Ethical Transparency*

Today's consumers are more informed and value-driven than previous generations. They demand clarity on the origin, environmental impact, and social footprint of the textiles they purchase. A 2021 McKinsey report revealed that 67% of fashion consumers consider using sustainable materials a vital purchasing factor [16]. Moreover, the rise of various digital platforms and social media activism has made it difficult for brands to hide behind opaque supply chains.

Movements such as *Fashion Revolution* have amplified calls for transparency through campaigns like #WhoMadeMyClothes, increasing awareness about worker exploitation, toxic dyeing practices, and wasteful production. Simultaneously, consumers are exploring and supporting slow fashion, thrift culture, and local artisan-based models, which inherently depend on more regenerative and ethical resource mixes.

This shift in values has prompted many brands to include fibre content disclosures, *Life Cycle Analysis (LCA)* reports, and third-party certifications (e.g., GOTS, OEKO-TEX, Cradle to Cradle). These measures are still evolving in terms of rigour and accessibility, indicating a growing alignment between consumer values and fibre innovation.

3.3 *Regulatory and Policy Frameworks*

Government policy plays an important role in shaping material choices across the value chain. With increasing global attention on climate action and sustainable development, several jurisdictions are now implementing or considering legally binding regulations for the fashion and textile sectors.

The *European Union's Strategy for Sustainable and Circular Textiles (2022)* is a landmark policy that mandates eco-design, durability, and repairability in clothing, as well as extended producer responsibility (EPR) schemes that hold brands accountable for post-consumer waste. France has already passed legislation requiring brands to disclose environmental impacts and prohibit the destruction of unsold stock (EPRS 2022).

In India, the *National Technical Textiles Mission (2020)* promotes innovation in functional and sustainable textiles, while public procurement policies are slowly incorporating green standards. The *U.S. FABRIC Act (2022)* aims to re-shore textile production and ensure compliance with labour and environmental laws. *International climate agreements,* such as the Paris Agreement and the upcoming COP28 deliberations, have placed pressure on textile-producing nations to align with net-zero targets. These policy pressures are accelerating investment in biodegradable, recyclable, and regenerative fibre systems, particularly those that demonstrate social inclusion and traceability.

3.4 Technological Innovation and Material Science

Innovation in fibre development, recycling technology, and biotechnology is enhancing the possibilities of the textile resource mix. New frontiers in *biofabrication* are enabling the creation of textiles from unexpected sources such as fungi, algae, agricultural waste, and even synthetic biology.

Some notable advancements include:

- Mycelium-based leather (e.g., Mylo™ by Bolt Threads)
- Bacterial cellulose textiles (e.g., by Modern Meadow)
- Recyclable polyester produced through chemical recycling (e.g., by Worn Again or Carbios)
- Agricultural waste fibres from banana, pineapple (Piñatex), and orange peel (Orange Fibre).

Simultaneously, AI and blockchain technologies are being deployed to optimise material selection, track supply chains, and measure environmental performance in real time. Brands like Lenzing and Kering now utilise fibre traceability platforms like TextileGenesis to validate sustainability claims and ensure authenticity across complex global supply chains. These innovations are transforming the landscape of raw materials and the economics of fashion. Regenerative and circular fibres are now viewed as strategic assets, driving competitive differentiation, brand reputation, and compliance with ESG standards (Table 1).

The transformation of the textile resource mix is the result of multifaceted systemic pressures and emerging opportunities. Climate realities constrain traditional fibre sources while driving the adoption of land-regenerating materials. Consumers, empowered by information and ethical frameworks, are demanding accountability and purpose from the products they purchase. Governments are tightening the regulatory noose, moving beyond voluntary sustainability to enforceable standards. Technology, meanwhile, enables unprecedented innovation from lab-grown fibres to AI-verified traceability.

Table 1 Key drivers of change in textile resource mix

Driver	Description	Impact on resource mix
Climate change & resource scarcity	Rising temperature, water stress, degraded soil, and reduced fossil fuel access	Shift to low-impact, regenerative, or alternative fibres
Consumer awareness & ethical demand	Demand for transparency, traceability, and value-driven consumption	Preference for certified, local, biodegradable, and recycled fibres
Policy & regulation	EU green policies, EPR mandates, trade compliance	Incentivizes biodegradable, durable, and traceable fibre systems
Technology & innovation	Biofabrication, recycling tech, AI, blockchain	Enables novel, waste-free, and regenerative fiber development

Together, these drivers fundamentally change how the industry perceives and procures its materials. As regenerative models evolve, the textile resource mix will increasingly prioritise planet-positive materials, decentralise production, and embed circularity as the norm rather than the exception.

4 Regenerative Natural Fibres and Their Eco System Benefits

The textile industry is increasingly looking beyond sustainability towards *regenerative practices* that actively restore ecosystems and strengthen the resilience of natural systems. At the heart of this transition is the shift towards regenerative natural fibres that are cultivated and processed in a manner that restores soil health, enhances biodiversity, sequesters carbon, and supports local communities. Unlike conventional practices that often degrade land and ecosystems, regenerative fibre systems are rooted in indigenous wisdom, ecological science, and holistic land stewardship.

This section examines four major categories of regenerative natural fibres: *cotton,animal-based fibres (such as wool and alpaca),bast fibres (hemp, flax, jute),* and *forest-based fibres (e.g., Tencel, bamboo, banana fibre).* Each fibre category not only offers unique functional and aesthetic qualities but also delivers tangible environmental benefits when produced in a regenerative manner.

4.1 Regenerative Cotton Farming

Cotton, the most widely used natural fibre globally, accounts for around 24% of global fibre production [3]. Traditionally, cotton farming has been notorious for its water consumption, pesticide use, and soil degradation. However, regenerative cotton farming is transforming this narrative by adopting techniques that rebuild soil fertility and increase resilience to climate variability.

Key regenerative practices in cotton cultivation include:

- **Crop rotation** to prevent pest buildup and restore nutrient cycles
- **Cover cropping** to reduce erosion and enhance organic matter
- **Compost and organic amendments** to enrich soil microbiota
- **Minimal or no tillage** to reduce carbon loss from soil.

These practices help mitigate environmental harm and sequester atmospheric carbon into the soil. The Rodale Institute's trials showed that regenerative organic systems could sequester over *1.5 tons of carbon per acre annually* [17]. Programs like the *Regenerative Organic Certification (ROC)* and *Fibershed's Climate Beneficial™ initiative* are supporting growers in this transition, particularly in India, the U.S., and parts of Africa.

Furthermore, regenerative cotton systems reduce input costs over time and promote agricultural sovereignty, particularly for smallholder farmers, who make up the majority of cotton growers in countries like India.

4.2 Wool and Animal-Based Fibres: Regenerative Grazing Systems

Wool, alpaca, yak, and other animal-based fibres have traditionally formed a symbiotic relationship with pastoral ecosystems. However, industrial-scale livestock farming has led to overgrazing, methane emissions, and biodiversity loss. In contrast, *regenerative grazing* offers a pathway to produce these fibres while improving land quality and ecosystem services.

Regenerative grazing involves *planned rotational grazing* that allows grasses to recover and prevents overgrazing. It also includes *mob grazing*, where animals are rotated intensively to mimic the movements of wild herds, and *Integrated land management,* which balances livestock numbers with the land's carrying capacity. These techniques promote deep root growth in grasses, enhancing carbon sequestration and increasing water retention in soil [18]. Research has shown that regenerative grazing systems can sequester up to 3 metric tons of CO_2 per hectare per year [19].

Organisations such as Savoury Institute and Fibershed are collaborating with ranchers and herders worldwide to implement these practices. Case studies from Patagonia in South America and merino wool producers in Australia show measurable improvements in soil organic carbon, biodiversity, and rural livelihoods. Moreover, regenerative grazing can support wildlife corridors, reduce desertification, and maintain cultural heritage for nomadic and indigenous communities reliant on pastoralism.

4.3 Bast Fibres: Hemp, Jute, Linen (Flax), and Ramie

Bast fibres are extracted from the outer layer of plants and have been used for millennia in clothing, ropes, and household textiles. Hemp, in particular, has re-emerged as a regenerative super-fibre due to its low input requirements and high carbon absorption.

These crops offer multiple ecological benefits:

- **Fast-growing and high-yielding**, requiring little or no pesticides
- **Deep-rooted systems** that prevent erosion and improve soil structure
- **Phytoremediation properties** of hemp, for example, can absorb heavy metals and detoxify polluted soil [20].

Jute and flax (linen) also require significantly less water than cotton and are grown in floodplains that do not compete with food crops. In regions like West Bengal (India) and Bangladesh, regenerative jute farming is being integrated with fish cultivation and rice paddies, showcasing a *multi-functional land-use model.* Bast fibres also biodegrade rapidly at the end of life, making them ideal candidates for *circular textile systems.* With increasing demand from luxury and wellness markets, regenerative bast fibres are also gaining traction in home textiles, fashion, and packaging.

4.4 Forest-Based Fibres: From Trees and Agricultural Residues

Forest-derived cellulose fibres, such as Tencel (Lyocell), bamboo viscose, and banana fibres, offer an alternative to both cotton and synthetic materials. When responsibly sourced, these fibres can be an integral part of *regenerative forestry models* that prioritise ecosystem restoration and fibre traceability.

Tencel is manufactured by Lenzing AG using sustainably harvested wood pulp from FSC-certified forests and a closed-loop solvent system, reclaiming over 99% of solvents used. Its production emits 50% less CO_2 and uses significantly less water compared to conventional cotton [21].

Banana fibre, widely used in Japan, India, and the Philippines, is made from agricultural waste (banana pseudo stems) and requires no additional land or water, making it highly sustainable. Similarly, bamboo fibres, when not chemically treated, grow without fertilisers and regenerate quickly, making them ideal for degraded lands.

Forest-based fibres support:

- **Agroforestry and permaculture systems**
- **Reforestation efforts** with native species
- **Revenue diversification** for small-scale farmers

However, these benefits depend on the traceability and harvesting methods employed. Unchecked bamboo and eucalyptus plantations can cause monoculture issues and water depletion. Therefore, certifications like the *Forest Stewardship Council (FSC)* and initiatives like *Canopy's Hot Button Ranking* help ensure that sourcing does not threaten ancient and endangered forests (Table 2).

While regenerative natural fibres hold immense promise, several *challenges* remain:

- **Scalability**: Regenerative practices often require labour-intensive methods, making them harder to scale rapidly.
- **Cost**: The initial transition period can be costlier due to training, certification, and land adjustments.
- **Verification and Certification**: Ensuring fibres are truly regenerative remains a challenge without robust monitoring tools and transparent data.

Table 2 Regenerative natural fibres and their ecosystem benefits

Fibre type	Key practices	Ecosystem benefits
Regenerative cotton	Crop rotation, cover cropping, compost application, minimal tillage	Soil carbon sequestration, improved soil health, water retention
Wool & animal fibres	Managed grazing, rotational pasture systems, biodiversity conservation	Increased soil fertility, carbon drawdown, wildlife habitat support
Bast fibres *(e.g., hemp, jute, flax)*	Low-water farming, fast renewability, no chemical inputs	Carbon capture, erosion control, less water and pesticide use
Forest-based fibres *(e.g., Tencel, bamboo, banana fibre)*	Sustainable forestry, closed-loop processing, reduced chemical use	Forest regeneration, lower impact on biodiversity, reduced deforestation

Despite these hurdles, regenerative models are increasingly proving economically viable, especially when considering ecosystem service payments, premium markets, and brand investment in ESG metrics. Regenerative natural fibres represent a critical solution in reimagining the textile industry's relationship with nature. They offer more than harm reduction; they actively restore ecological function, sequester carbon, enrich biodiversity, and uplift rural communities. By adopting fibre choices rooted in regeneration, the textile industry can move beyond sustainability rhetoric to real environmental repair.

Embedding regenerative fibres into mainstream production will require *collaborative partnerships* across farming communities, brands, certifiers, scientists, and policymakers. As climate change accelerates and consumer expectations shift, regenerative natural fibres offer a path forward that is both timeless and future-ready.

5 Circular and Recycled Materials: Closing the Loop

As the textile and fashion industry grapples with unprecedented environmental challenges, the integration of circular and recycled materials into production systems has emerged as a vital strategy for sustainability. The traditional linear model "take, make, dispose" has proven unsustainable, with over 92 million tonnes of textile waste generated globally every year [1]. A regenerative future for textiles necessitates a radical redesign of this model into a circular system where materials are kept in use for as long as possible, their value is maintained through repair and reuse, and waste is minimised through effective recycling. Circularity not only conserves raw resources but also minimises carbon emissions, water usage, and toxic waste.

In the context of regenerative textile systems, closing the loop involves the extensive use of recycled materials, both pre-consumer (factory offcuts) and post-consumer (discarded garments) and the implementation of textile-to-textile recycling, biowaste

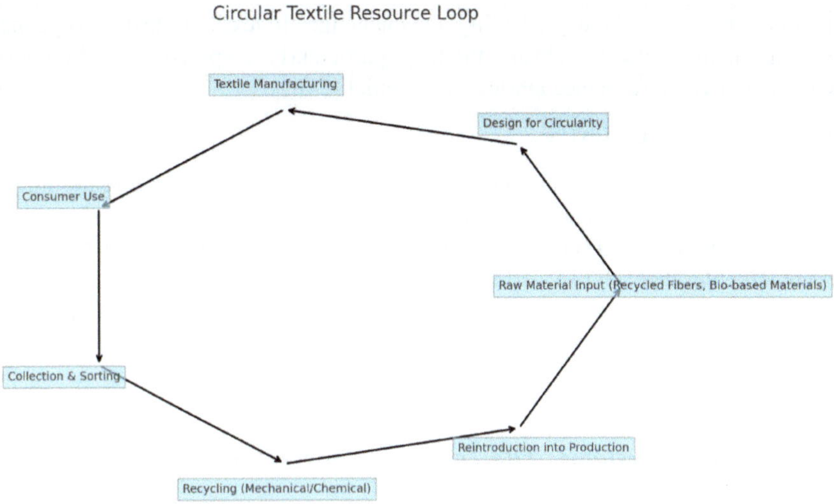

Fig. 2 Circular textile resource loop

upcycling, and the use of bio-based synthetics. These methods represent an evolution from mere sustainability to resource regeneration, where textile waste becomes a raw material for new value chains.

Here is a visual representation of the *Circular Textile Resource Loop,* illustrating how materials flow through a circular system, starting from recycled or bio-based inputs to reintroduction after consumer use and recycling (Fig. 2).

5.1 Types of Circular and Recycled Materials

A. Textile-To-Textile Recycled Fibres

Recycling textiles back into fibres for reuse is one of the most impactful interventions in reducing dependency on virgin materials. Mechanical recycling involves shredding fabrics and re-spinning the fibres, typically used for wool, denim, and cotton, though fibre quality degrades with each cycle [22]. Chemical recycling, a more advanced method, breaks fibres down into their molecular components to regenerate high-quality fibres, allowing for near-infinite recyclability in some cases [23]. Companies like Renewcell and Worn Again Technologies are pioneers in this domain, recycling blended textiles and cellulose-rich waste into new fibres like Circulose®.

b. Recycled Polyester (rPET)

rpet, derived from post-consumer plastic bottles and used garments, is among the most common recycled synthetic fibres. While it diverts plastic waste from landfills and reduces dependence on fossil-based virgin polyester, its environmental benefit is limited by its non-biodegradability and challenges in textile-to-textile recycling. Still, it remains critical in closing the loop, particularly in sportswear and outdoor apparel, where performance fabrics are essential [24].

c. Biowaste-Derived Textiles

Innovative bio-based materials made from agricultural or food waste are entering the textile ecosystem as regenerative alternatives. Piñatex® (pineapple leaves), Orange Fiber® (citrus peel), and Bananatex® (banana plant stalks) represent a new frontier in circular textiles that valorise waste streams while minimising ecological harm. These materials embody cradle-to-cradle thinking, transforming what was once discarded into high-value textile inputs [25].

5.2 Circular Systems: The Textile Resource Loop

Circular systems depend on the seamless integration of material reuse, consumer engagement, and infrastructure for take-back, sorting, and recycling. The previously shown diagram Fig. 2 illustrates the circular textile resource loop. This model is based on feedback loops where discarded materials are reintroduced into the production cycle. Each phase, such as design, use, recovery, and recycling, plays a critical role in maintaining material value and reducing environmental impact.

5.3 Design for Circularity

Circularity begins at the design phase. Products designed for longevity, reparability, disassembly, and mono-materiality simplify recycling and reduce waste. For instance, modular garments and mono-fibre textiles (e.g., 100% polyester or 100% cotton) eliminate the complications posed by blended fabrics, which are notoriously difficult to separate and recycle. Design strategies like cradle-to-cradle certification and the Ellen MacArthur Foundation's "Circular Design for Fashion" framework guide brands toward regenerative product development [26].

5.4 Blended Material Challenges

Despite the promise of recycling, blended textiles especially those combining natural and synthetic fibres, such as cotton-polyester blends, pose significant challenges. Mechanical recycling typically requires homogeneous input, and even advanced chemical processes are often limited by contaminants, dyes, and fabric complexity [22]. This has led to the development of technologies like hydrothermal separation and enzymatic degradation that are capable of selectively extracting components from blended materials.

5.5 Consumer Role and Post-Consumer Waste Collection

Circularity cannot be achieved without the active participation of consumers. Efficient systems for garment take-back, sorting, and resale depend on informed consumers. Brands like Patagonia and Eileen Fisher have implemented successful product return and resale programs, while European policies such as Extended Producer Responsibility (EPR) mandates require producers to manage post-consumer waste [27].

Digital tools such as QR code tags and blockchain are being used to track a garment's lifecycle, ensuring transparency and aiding in efficient sorting and recycling. Consumer behaviour also impacts garment life; extending use by just nine months can reduce environmental impact by 20–30% [24].

5.6 Re-Commerce and Resale

Second-hand fashion and re-commerce platforms are experiencing exponential growth, signalling a shift in consumer mindset. ThredUp's 2023 Resale Report projects that the global resale market will double in value by 2027. By extending the lifespan of garments through reuse and resale, brands can reduce demand for virgin materials and promote circular business models.

Brands like Levi's, The North Face, and H&M are integrating resale platforms into their business ecosystems, often supported by third-party services like Trove or Reflaunt. These platforms digitise inventory and customer engagement, facilitating seamless product returns and resales.

5.7 Policy and Industry Collaboration

Circularity requires systemic transformation, which is only possible through cross-sectoral collaboration and policy intervention. Governmental policies like the EU Strategy for Sustainable and Circular Textiles and France's anti-waste law for a circular economy mandate recycling, prohibit unsold inventory destruction, and encourage EPR. These frameworks push brands to adopt circular practices not as a marketing strategy but as a compliance necessity [27].

Collaborative initiatives such as the Circular Fashion Partnership and the Global Fashion Agenda's Circularity ID support industry alignment and knowledge sharing.

These coalitions promote innovation in sorting technology, recycling infrastructure, and the development of new materials.

5.8 *Towards Regenerative Circularity*

Circularity alone may not be enough if it masks harmful industrial practices. Regenerative circularity emphasises not just closed-loop production but also the regeneration of ecosystems, livelihoods, and social equity. For instance, sourcing recycled cotton from regenerative farms or reintroducing bio-waste from agroforestry systems contributes to a more holistic regenerative vision.

Brands can adopt circular models that prioritise degrowth, producing less but better, while ensuring community engagement and fair labour practices. When integrated with regenerative agriculture, decentralised recycling, and local economies, circular materials can redefine the textile value chain in alignment with planetary boundaries.

Circular and recycled materials offer a compelling pathway towards closing the loop in the textile industry. From rPET and textile-to-textile recycling to biowaste valorisation and resale ecosystems, the textile resource mix embraces circularity not just as a sustainability goal but as a foundation for long-term ecological resilience.

6 Bioengineered & Alternative Fibres: The Future of Resource Innovation

Traditional fibre sources, both natural and synthetic, are increasingly being re-evaluated for their high ecological footprint, intensive resource consumption, and challenges related to biodegradability. In response, a new generation of fibres is emerging that transcends the limitations of existing materials. Bioengineered and alternative fibres represent not just sustainable options but regenerative solutions that align with the future of a circular, climate-positive textile economy.

6.1 *Understanding Bioengineered and Alternative Fibres*

Bioengineered fibres refer to materials that are developed using biotechnology, microbial fermentation, or cellular agriculture to replicate or enhance the properties of natural fibres without the associated ecological burden. These innovations are not simply substitutes; they are purpose-built from the molecular level to meet both environmental and performance needs. Alternative fibres, meanwhile, include those derived from unconventional resources such as agricultural waste, fungi, algae, or plant proteins that are typically overlooked in mainstream textile production.

Together, these fibre types challenge existing paradigms of resource extraction, waste generation, and carbon emissions. Their ability to reduce dependence on water-intensive crops, fossil-fuel synthetics, and animal-based materials positions them as vital contributors to a regenerative textile future.

6.2 Categories and Leading Innovations

1. Mycelium-Based Textiles

Mycelium, the root-like structure of fungi, has gained significant traction as a scalable bio-fabricated material. Products such as Mylo™, developed by Bolt Threads, replicate the feel and appearance of leather without involving animals or toxic tanning processes. Mycelium is grown on agricultural waste, requiring minimal water, and decomposes naturally at the end of its life cycle [28]. Compared to animal leather, it uses significantly fewer resources and emits a fraction of the greenhouse gases.

2. Bacterial Cellulose

Produced through the fermentation process using bacteria like *Acetobacter xylinum*, bacterial cellulose yields thin, flexible films that can be fashioned into wearable materials. Companies like BioCouture™ have demonstrated bacterial cellulose's potential as an alternative to plastic-based and animal-derived materials. It requires no land use or pesticides and can be composted after use [29]. With innovations in microbial engineering, bacterial cellulose can be cultivated in tailored forms, thereby reducing waste during cutting.

3. Lab-Grown Silk and Wool

Biotechnology firms such as AMSilk and Spiber have pioneered the synthesis of spider silk proteins via genetically engineered yeast and bacteria. These proteins are spun into fibres that possess remarkable tensile strength, elasticity, and biodegradability. Unlike conventional silk, no silkworms are harmed or raised under industrial conditions, and the energy demand is comparatively lower [30].

Similarly, lab-grown wool developed through cellular agriculture by companies like VitroLabs replicates the follicular structure of sheep wool. This allows for fine-tuned fibre attributes while bypassing land-use, animal ethics, and methane emissions associated with conventional wool farming.

4. Agricultural Waste Fibres

A growing segment of alternative fibres is derived from post-harvest agricultural waste, which is otherwise burned or discarded. These include:

- **Piñatex®** from pineapple leaf fibre
- **Orange Fiber®** from citrus waste

- **Bananatex®** from banana plant stems.

These fibres exemplify circular design, valorising waste streams while requiring no additional water, land, or fertiliser [31]. Brands like Hugo Boss and H&M have incorporated such fibres into their sustainable collections, signalling growing commercial acceptance.

5. Protein-Based Plant Fibres

Derived from soy, peas, and grape skins, protein-based fibres offer an alternative to petroleum-based synthetics and animal fibres. Products like SoySilk™ and VEGEA use renewable agricultural by-products to produce soft, durable textiles without resorting to harsh chemical processing. These fibres are fully biodegradable and can be integrated into composting systems, aiding nutrient cycling in regenerative agricultural models [32].

6. Algae-Based Fibres

Algae, including microalgae and seaweed, are the most promising bio-resources for fibre production. They grow rapidly, sequester CO_2, and can thrive in non-arable environments. Companies such as AlgaLife and Algiknit convert algae biomass into protein- and cellulose-rich fibres used for knitting, weaving, and non-woven applications. These fibres are inherently antimicrobial and hypoallergenic, and their production requires no freshwater or pesticides [33] (Table 3).

Table 3 Innovative bioengineered & alternative fibres for a regenerative future

Fibre type	Source material	Sustainability benefits	Commercial examples
Mycelium based textiles	Fungi (Mycelium)	Bio degradable, Low Energy production	Mylo™ by Bolt Threads
Bacterial cellulose	Bacteria (e.g. Acetobacter xylinum)	Water efficient, compostable, biodegradable	Bio Culture™
Lab-grown silk (spider silk)	Yeast-engineered spider silk proteins	No animal harm, light weight, strong	Spiber, AMSilk
Lab grown wool	Cell cultures mimicking sheep follicles	Eliminates animal farming	Vitrolabs
Agricultural waste fibres	Orange peels, Pineapple leaves, Banana Stems	Utilises food and Agri waste, Reduces landfill burden	Piñatex®, Bananatex®, Orange Fiber®
Protein based fibres	Soy, Peas, Corn Proteins	Renewable, Bio degradable, Low GHG emissions	SoySilk™, VEGEA
Algae based fibres	Algae, Sea Weed	Carbon Absorbing, Fast Growing Biomass	AlgaLife, Algiknit

6.3 Environmental and Systemic Benefits

The bioengineered and alternative fibres detailed above collectively offer significant environmental benefits such as.

- **Biodegradability**: Most of these fibres decompose naturally, leaving no toxic residues or microplastics.
- **Resource Efficiency**: They often utilise waste streams, require less water and energy, and minimise the use of harmful chemicals.
- **Waste Valorisation**: By using food waste or biomass, they close the loop and reduce the burden on landfills.
- **Animal Welfare**: No animals are harmed or exploited, aligning with cruelty-free and vegan fashion ideals.
- **Carbon Footprint Reduction**: Many fibres sequester carbon during growth and avoid fossil-fuel-based production.

These attributes make them ideally suited for inclusion in regenerative systems that restore ecosystems, foster biodiversity, and regenerate soil.

6.4 Challenges and Implementation Barriers

Despite their immense promise, these innovative fibres face considerable challenges:

- **High Production Costs**: Bioengineering technologies remain expensive and are largely confined to niche markets.
- **Scale Limitations**: Most companies are still in the pilot or early commercial stages, which limits global adoption.
- **Consumer Awareness**: End-users may lack familiarity or trust in unfamiliar materials.
- **Regulatory Ambiguity**: Novel materials often fall outside established textile certification or trade regulations.
- **Performance Consistency**: Achieving industrial-level strength, washability, and colourfastness remains a significant R&D challenge.

Therefore, collaborations between biotech firms, fashion brands, governments, and academic institutions are essential to enable large-scale transitions.

6.5 Pathways to Integration in Regenerative Textile Systems

For these fibres to truly drive the transition toward regenerative industry, they must be integrated into broader systems thinking, including:

- **Design for Circularity**: Fibres should be compatible with recycling systems, composting, or remanufacturing.
- **Supply Chain Transparency**: Blockchain and other digital tools can ensure traceability of bioengineered inputs.
- **Farmer and Artisan Engagement**: For agricultural waste fibres, partnerships with rural communities can help decentralise production.
- **Consumer Co-education**: Material storytelling and digital labelling can empower consumers to make informed purchasing decisions.

Bioengineered and alternative fibres mark a paradigm shift in how we envision the raw materials of textiles. More than just sustainable, they offer pathways toward regenerative design, where materials do less harm and actively contribute to environmental restoration and social value. The integration of innovations in biotech, agriculture, and circular economy principles, these fibres represent a viable and necessary future for the industry.

7 Enablers of a Regenerative Textile Industry

The transition from a resource-extractive, linear textile industry to a regenerative, restorative model requires more than just material innovation. It calls for systemic shifts across governance, business models, infrastructure, and consumer behaviour. *Regenerative design*, by definition, moves beyond reducing harm to creating positive ecological and social impact. This vision demands a collaborative ecosystem enabled by new tools, policies, and mindsets.

7.1 Regenerative Business Models

One of the foundational enablers is the emergence of regenerative business models that challenge the conventional *'take-make-waste'* approach. Regenerative models embed circular economy principles but go a step further by focusing on restoring ecosystems, empowering communities, and enhancing long-term resilience. For example, Patagonia's partnership with regenerative organic farms to produce cotton demonstrates how businesses can invest in supply chains that actively rebuild soil health, increase biodiversity, and sequester carbon [34]. Similarly, India's Fibreshed initiatives are creating decentralised, bioregional textile economies that reduce emissions, shorten supply chains, and foster community ownership [35]. Key elements of regenerative business models include:

- Localised and circular supply chains
- Extended product lifecycles through repair, reuse, and resale
- Revenue models based on services (e.g., rental, subscription) rather than volume sales

- Equitable value distribution for growers, artisans, and workers.

These models redefine value not only in terms of profit but also in terms of ecosystem restoration and social equity.

7.2 *Material Traceability and Digital Transparency*

The complexity of global textile supply chains often obscures materials' origin, composition, and ethical footprint. Traceability technologies, such as blockchain, QR codes, radio-frequency identification (RFID), and digital product passports, are increasingly being used to map the entire lifecycle of textiles. Blockchain platforms such as TextileGenesis and Provenance provide immutable records of every transaction in a material's journey from farm to finish. These technologies enable brands, regulators, and consumers to verify claims such as "organic," "regenerative," or "recycled" [36].

Digital transparency tools:

- Improve accountability across the supply chain
- Encourage ethical sourcing by exposing harmful practices
- Enable informed consumer choices
- Simplify compliance with emerging sustainability regulations.

The European Union's proposed *Ecodesign for Sustainable Products Regulation (ESPR)* will require digital product passports for textiles by 2030, making traceability not just a value-add but a regulatory necessity (European Commission, 2022).

7.3 *Extended Producer Responsibility (EPR) Frameworks*

Extended Producer Responsibility (EPR) policies require producers to bear responsibility for the entire lifecycle of their products, particularly for post-consumer disposal and recycling. EPR schemes are pivotal in moving the industry from linear to circular, and eventually to regenerative systems. EPR can incentivise:

- Design for durability, recyclability, and biodegradability
- Investment in take-back, reuse, and recycling infrastructure
- Reduction of overproduction and fast fashion cycles.

For example, France has implemented a robust EPR scheme under Refashion, where brands contribute to the cost of collecting, sorting, and recycling post-consumer textiles. As of 2023, over 500 million garments have been processed under this system [37]. India, too, is in the process of drafting EPR policies for textiles, which would hold brands accountable for waste management and encourage upstream design innovations.

7.4 Financial and Policy Incentives

Policy support and financial innovation are critical enablers for scaling regenerative practices in the textile industry. Governments and multilateral institutions are beginning to offer incentives for environmentally positive textile practices. Examples include:

- Subsidies for regenerative agriculture and organic cotton cultivation
- Tax breaks for companies investing in circular technologies
- Green bonds and ESG funds for sustainable textile start-ups
- Import tariffs favouring recycled or low-impact textiles.

The Global Fashion Agenda and the UNFCCC's Fashion Industry Charter for Climate Action also recommend including regenerative practices in national climate strategies [38]. Policy harmonisation across global markets can further reduce barriers to adoption and standardise regenerative criteria.

7.5 Regenerative Infrastructure and Innovation Hubs

Infrastructure forms the backbone of the regenerative textile economy. Without investments in localised spinning, weaving, composting, and recycling facilities, regenerative materials cannot reach scale or remain economically viable.

In countries like the Netherlands, circular textile hubs are being developed to integrate post-consumer collection, fibre sorting, and mechanical or chemical recycling [39]. In India, clusters like Kutch and Odisha are experimenting with bioregional production systems that integrate natural dyeing, handweaving, and farm-based fibre sourcing. Innovation hubs and incubators also play a vital role in nurturing regenerative solutions. Programs like Fashion for Good and Ashoka's Fabric of Change provide funding, mentorship, and networks for start-ups that are creating regenerative dyes, circular garments, and AI-powered waste management systems.

7.6 Collaboration Across the Value Chain

Multi-stakeholder collaborations across brands, producers, researchers, policymakers, and consumers are essential. Cross-sectoral partnerships enable knowledge exchange, risk sharing, and alignment of standards. For instance, the Regenerative Cotton Fund in the U.S., initiated by the Soil Health Institute and supported by major brands, aims to transition 1 million acres to regenerative practices by 2026 [40]. The fund brings together farmers, NGOs, and apparel companies to co-design solutions. Such collaborations:

- Accelerate innovation and standardisation

- Support capacity building for smallholders and artisans
- Harmonise definitions and metrics for regeneration
- Build shared ownership in regenerative goals.

7.7 Consumer Participation and Cultural Shifts

Regeneration requires a cultural shift in consumption patterns. The success of regenerative textiles relies on active consumer participation through mindful purchasing, proper care, and responsible disposal.

Educational initiatives, transparent storytelling, and experiential retail can encourage consumers to:

- Choose fewer but higher-quality garments
- Support regenerative brands and practices
- Participate in repair, swap, and resale programs
- Understand the environmental and social impact of their wardrobe.

Campaigns like *#WhoMadeMyClothes* and *"Loved Clothes Last"* by Fashion Revolution are powerful in building consumer consciousness [41]. When consumers demand regenerative values, the market is compelled to respond.

The pathway to a regenerative textile industry is not paved by materials alone but by the convergence of enabling systems, policies, tools, and mindsets. Regenerative business models, material traceability, supportive policy frameworks, infrastructural development, cross-sector collaboration, and informed consumer behaviour are the essential scaffolds of this future. These enablers help translate regenerative principles into real-world outcomes: healthy soils, equitable livelihoods, circular economies, and thriving ecosystems. The industry must now leverage these enablers not as peripheral add-ons but as central pillars of a just, regenerative transition.

8 The Road Ahead: A Vision for Regenerative Textiles

As the textile and fashion industries confront the dual crises of ecological degradation and social inequity, the path forward demands more than sustainability. It demands regeneration. Regenerative textiles envision an industry that minimises harm, actively heals ecosystems, empowers communities, and redefines value. The future of textile resource mixes lies in multi-dimensional strategies that go beyond material substitutions to systemic transformations across policy, culture, economics, and technology.

This chapter outlines a practical and visionary roadmap for enabling regenerative textile futures. It considers how global collaboration, policy coherence, bioregional design, and cultural shifts can accelerate the integration of regenerative practices into the heart of the textile economy.

8.1 *From Degenerative to Regenerative Thinking*

The starting point in envisioning the road ahead is a philosophical and epistemological shift from *linear*, extractive thinking to regenerative mindsets. Where traditional models ask, 'How can we reduce impact?', regenerative frameworks ask, 'How can we restore life systems?'.

Daniel Wahl (2016) defines regenerative design as "creating the conditions conducive for life." In textiles, this implies creating material systems that replenish soils, increase biodiversity, restore water cycles, and support dignified livelihoods. This reframing challenges the current norm of efficiency-based models that focus on doing less harm rather than creating more good. Designers and supply chain actors must be reoriented to see themselves as stewards of ecosystems rather than merely producers or profit generators. The textile industry of the future must align with the logic of natural systems: cyclical, diverse, place-based, and interdependent.

8.2 *Expanding the Paradigm with Circularity*

The circular economy has been widely promoted as the pathway to reducing waste and decoupling resource use from growth. However, circularity alone does not guarantee ecosystem recovery or social justice. In contrast, regenerative systems are built on the recognition that human activity can be a force for healing rather than harm [9]. Where circularity seeks to minimise negative outputs, regeneration emphasises positive contributions, such as replenishing soil fertility through regenerative cotton farming, or enhancing biodiversity via rotational grazing for wool production. Thus, the regenerative textile industry of the future must prioritise:

- **Holistic value creation**: Integrating economic, ecological, and cultural capital.
- **Systemic design**: Considering textiles' whole lifecycle, social networks, and ecological contexts.
- **Restorative feedback loops**: Encouraging practices that restore natural cycles, not merely reuse inputs.

8.3 *Regenerative Thinking in Design and Innovation*

Design is one of the most powerful leverage points in transforming the textile system. The future of regenerative textiles must embrace "design for regeneration" as a principle, incorporating biomimicry, indigenous knowledge, and long-term thinking into material and product development. *Biomimicry*, for instance, can inspire textiles that mimic the self-cleaning properties of lotus leaves or the cooling efficiency of termite mounds [42]. Meanwhile, *indigenous craft epistemologies* such as zero-waste weaving techniques or natural dyeing with seasonal botanicals offer time-tested regenerative insights rooted in place-based knowledge [43]. Key innovations likely to shape the regenerative future include:

- Biofabricated textiles with biodegradable life cycles.
- Modular, upgradable clothing designed for longevity.
- Decentralised, community-owned production systems.
- Materials that actively detoxify air or enrich soil at end-of-life.

8.4 Re-Embedding Textiles in Ecosystems

Regenerative textiles are not commodities extracted from nature; they are extensions of the living systems from which they come. In future models, textile production must become bioregionally attuned, rooted in the capacities, rhythms, and limits of local ecosystems.

This means rethinking not only *what* materials we use but *where* and *how* they are produced:

- **Fibersheds** are bioregional systems that connect fibre, dyes, and labour within a shared geography. They demonstrate how locally grown and processed textiles can reduce emissions and enhance resilience [44].
- **Agroecological integration** will link textile crops to food production, water cycles, and soil regeneration goals.
- **Watershed-sensitive dyeing** will prioritise non-toxic, closed-loop processes that protect aquatic ecosystems.

Such localised systems will not only reduce ecological footprints but also reclaim cultural relationships between people and materials.

8.5 Bridging Technological and Traditional Knowledge

A regenerative vision must reconcile the dual imperatives of technological progress and cultural preservation. It is not about rejecting innovation but about aligning it with planetary boundaries and ancestral wisdom. For instance, *AI-powered sorting systems* can enhance the efficiency of textile recycling, while *blockchain* ensures transparency and traceability (Ellen MacArthur Foundation, 2021). At the same time, handloom weaving communities in India, Peru, and Morocco offer low-energy, high-skill solutions that foster dignified livelihoods and cultural continuity [45]. In the future, regenerative systems will:

- Combine high-tech with high-touch craftsmanship.
- Enable data-driven yet culturally rooted supply chains.
- Support transdisciplinary education blending design, ecology, and indigenous knowledge.

This convergence allows for scalable solutions that remain contextual, ethical, and regenerative.

8.6 Equitable Participation and Economic Justice

A regenerative textile system must be both socially healing and environmentally restorative. Equity, inclusion, and justice are non-negotiable pillars. The transition cannot simply greenwash extractive models or offload responsibility onto marginalised communities. Future regenerative economies must:

- Ensure fair wages and safe working conditions throughout supply chains.
- Recognise and remunerate the intellectual contributions of indigenous and artisan communities.
- Decentralise ownership through cooperatives, community enterprises, and participatory governance mechanisms.

Regenerative justice encompasses *cultural restitution*, which involves restoring the stories, names, and recognition to heritage textiles that global brands have appropriated without acknowledgement or compensation [46]. Only with equity at the core can regenerative systems avoid replicating colonial and capitalist patterns of exploitation.

8.7 Measuring What Matters

Moving toward a regenerative textile economy will also require shifting what we measure. Traditional metrics such as GDP or productivity fail to capture ecological regeneration or social well-being. Emerging frameworks such as Doughnut Economics [8] or the *Genuine Progress Indicator (GPI)* offer alternative models that value ecological health, community cohesion, and cultural richness.

Future textile impact assessments must include:

- **Soil health indexes** for fibre agriculture.
- **Biodiversity metrics** in grazing and dyeing practices.
- **Social return on investment (SROI)** in artisan livelihoods.
- **Carbon sequestration potential** in regenerative farms.

These tools will help organisations make informed, values-aligned decisions that prioritise long-term resilience over short-term profits.

8.8 Policy and Global Governance for Regeneration

While regenerative practices often emerge at grassroots or regional levels, scaling their impact requires supportive global policy frameworks. Governments and international bodies must play an active role in mainstreaming regenerative textiles. Future policy priorities include:

- Mandating regenerative sourcing for public procurement and fashion brands.
- Incentivising regenerative agriculture and textile recycling through subsidies and tax relief.
- Imposing tariffs on non-regenerative, high-impact imports.
- Funding research and capacity-building programs in regenerative design and production.

Moreover, global cooperation is vital. Regenerative transitions must be context-sensitive, avoiding one-size-fits-all mandates and allowing for diverse, place-based adaptations.

8.9 The Role of Education and Cultural Transformation

No regenerative future can emerge without a fundamental shift in values, worldviews, and cultural narratives. Formal and informal education plays a crucial role in cultivating regenerative mindsets. Institutions must revise curricula to integrate:

- Ecological literacy and systems thinking.
- Indigenous textile traditions and regenerative philosophies.
- Applied skills in composting, dyeing, upcycling, and biomaterials.
- Storytelling that nurtures empathy, interconnectedness, and imagination.

Creative platforms such as regenerative fashion festivals, craft residencies, and immersive exhibitions can engage the public in embodied, joyful, and hopeful textile futures. Ultimately, regeneration is not merely technical; it is spiritual, emotional, and relational. Regenerative textiles are not a utopian dream. They are a necessary evolution of how we design, make, and relate to materials. By embedding circularity with restoration, innovation with tradition, and production with care, regenerative systems offer a path to reconciling the industry with planetary boundaries and human dignity. The road ahead is long, nonlinear, and uncertain. However, it is also rich in possibilities. In imagining regenerative futures, we must hold a dual vision; one that is radically hopeful and rigorously grounded. It is in this balance that transformation becomes not only desirable but achievable.

Conclusion: A Paradigm Shift in Progress from Extraction to Regeneration

The future of textiles hinges not merely on sustainability, but on a broader and more profound transformation that requires a fundamental rethinking of how we source, use, and dispose of resources. As this chapter has explored, the textile industry's current model, long reliant on extractive and linear resource mixes, is no longer viable in the face of mounting ecological and social crises. The path forward must transcend incremental fixes and instead embrace regenerative thinking as a blueprint for systemic renewal.

In this context, regeneration is not just a material consideration; it is a design principle, a systems framework, and an ethical imperative. It challenges the industry to ask not only how to reduce harm, but also how to actively contribute to the restoration of ecosystems, cultures, and communities. This requires looking beyond fibre content or water usage statistics and toward a more holistic understanding of value, rooted in relationships between people, land, and materials.

What becomes clear is that resource mixes of the future will be dynamic, context-driven, and circular by design. They will draw from a wide array of sources—regenerative natural fibres, agricultural residues, lab-grown biofibres, and recycled content—all selected not merely for their functionality, but for their ability to restore life cycles, reduce dependency on virgin extraction, and fit harmoniously within ecological boundaries. The shift to regenerative resource mixes is not just a technological or environmental issue; it is a cultural and economic transition. It demands dismantling outdated assumptions, such as growth without limits, fast fashion as a norm, or the undervaluation of artisan labour. It also calls for a redistribution of power and recognition, particularly toward communities whose knowledge systems and practices have long embodied regenerative principles but have been marginalised in mainstream industrial narratives.

Moving forward, success will not be measured solely by the number of recycled garments or carbon credits achieved, but by the vibrancy of local textile ecosystems, the resilience of artisan livelihoods, and the health of soils, rivers, and biodiversity that support textile production. Policy frameworks, educational institutions, and consumer behaviour must align to support this vision, fostering a regenerative literacy across every stakeholder in the textile value chain. In essence, the regenerative textile industry of tomorrow is not an abstract ideal. It is already emerging through grassroots innovations, indigenous practices, and science-driven experimentation. It is found in the slow-grown cotton fields of farmers practising agroecology, in the looms of craftspeople working with zero waste, in the labs creating new materials that mimic nature, and in the hands of consumers choosing to repair, repurpose, and reconnect with what they wear.

To accelerate this transformation, the textile industry must act decisively and collaboratively. It must replace linearity with circularity, extraction with regeneration, and opacity with transparency. The future is not only about the materials we choose but also the systems we design, the values we uphold, and the futures we collectively imagine. Reimagining textile resource mixes for a regenerative industry is a profound opportunity not only to mitigate the textile industry's impact but also to reposition it as a powerful force for planetary healing and social renewal.

References

1. Ellen MacArthur Foundation (2017) A new textiles economy: redesigning fashion's future. https://ellenmacarthurfoundation.org/a-new-textiles-economy
2. Fletcher K, Tham M (2019) Earth logic: fashion action research plan. The JJF
3. Textile Exchange (2023) Preferred fibre and materials market report 2023. https://textileexchange.org
4. WWF (2021) Cotton: the fabric of our lives is straining our planet. https://www.worldwildlife.org
5. LaCanne CE, Lundgren JG (2018) Regenerative agriculture: merging farming and natural resource conservation profitably. PeerJ 6:e4428. https://doi.org/10.7717/peerj.4428
6. Fibershed (2022) Regenerating our textile system. https://fibershed.org
7. Lee A, Appels F, D'Espaux L, Egli N (2020) Emerging mycelium materials for design and architecture: current progress and future opportunities. Mater Today Sustain 7–8:100024
8. Raworth K (2017) Doughnut economics: seven ways to think like a 21st-century economist. Chelsea Green Publishing
9. Mang P, Reed B (2012) Designing for regeneration: principles for holistic sustainability. J Clean Prod 37:78–85. https://doi.org/10.1016/j.jclepro.2012.06.007
10. Pradhan S (2020) Interwoven lives: handloom, identity, and community. Unpublished manuscript
11. Savory Institute (2021) Holistic management and regenerative grazing. https://www.savory.global
12. Whyte KP (2018) Indigenous science (fiction) for the Anthropocene: ancestral dystopias and fantasies of climate change crises. Environ Plan E: Nat Space 1(1–2):224–242
13. UNEP (2019) Sustainability and circularity in the textile value chain: global stocktaking. United Nations environment Programme. https://www.unep.org
14. Chapagain AK, Hoekstra AY, Savenije HHG, Gautam R (2006) The water footprint of cotton consumption. Value of water research report series no. 18. UNESCO-IHE institute for water education
15. Boucher J, Friot D (2017) Primary microplastics in the oceans: a global evaluation of sources. IUCN. https://portals.iucn.org
16. McKinsey & Company (2021) The state of fashion 2021. Business of Fashion and McKinsey
17. Rodale Institute (2014) Regenerative organic agriculture and climate change: a down-to-earth solution to global warming. https://rodaleinstitute.org
18. Teague R, Apfelbaum S, Lal R, Kreuter U, Rowntree J, Davies C, Conser R, Rasmussen M, Hatfield J, Wang T, Wang J, Byck P (2016) The role of ruminants in reducing agriculture's carbon footprint in North America. J Soil Water Conserv 71(2):156–164
19. White RE, Davidson B, Lam SK, Chen D (2013) Can soil carbon sequestration in grazed pastures help counter climate change? Animal 7(s2):262–272
20. Shi G, Wang Y, Zhang S, Wu Y (2012) Phytoremediation potential of hemp (Cannabis sativa L.) and its implications for contaminated soils. J Environ Sci 24(6):1154–1160
21. Lenzing Group (2022) Sustainability report. https://www.lenzing.com
22. Sandin G, Peters GM (2018) Environmental impact of textile reuse and recycling—a review. J Clean Prod 184:353–365. https://doi.org/10.1016/j.jclepro.2018.02.266
23. Textile Exchange (2021) Preferred fibre & materials market report 2021. https://textileexchange.org
24. WRAP (2020) Valuing our clothes: the cost of UK fashion. https://wrap.org.uk/resources/report/valuing-our-clothes-the-cost-of-uk-fashion
25. de Silva S, Wang X, Rahman MM (2023) Agricultural waste-derived textile materials: sustainable options for the fashion industry. J Clean Prod 410:137416. https://doi.org/10.1016/j.jclepro.2023.137416
26. Ellen MacArthur Foundation (2021) A new textiles economy: redesigning fashion's future. https://ellenmacarthurfoundation.org

27. European Commission (2022) Proposal for a regulation on ecodesign for sustainable products. https://environment.ec.europa.eu
28. Jones J, Patel S, Lin D (2021) Mycelium: the future of biodegradable material in textiles. J Sustain Des 9(2):103–115
29. Yusuf M, Shabbir M, Mohammad F (2020) Bio-fabricated bacterial cellulose in textiles. J Nat Fibers 17(12):1653–1668
30. Zhao W, Liu X, Chien Y (2022) Genetically engineered spider silk in textile applications. Biotechnol Adv 60:108013
31. Hethorn J, Ulasewicz C (2022) Sustainable fashion: what's next? A conversation about issues, practices, and possibilities (3rd ed.). Fairchild Books
32. Patagonia (2021) Soy-based textiles: the next plant frontier? https://www.patagonia.com
33. Venkatesh R, Marimuthu R, Chauhan S (2023) Algae-derived fibres: innovations in bio-based textiles. Int J Fash Technol 14(1):56–69
34. Patagonia (2020) Regenerative organic certified cotton. https://www.patagonia.com/stories
35. Mukherjee R (2022) Bioregional textile economies and the role of Indian fibresheds. J Sustain Fash 7(3):76–91
36. Choi TM, Ng SF (2020) Sustainable fashion supply chain management: a review of emerging technologies and tools. Sustainability 12(24):10512. https://doi.org/10.3390/su122410512
37. Refashion (2023) Annual report 2022: French EPR for textiles. https://refashion.fr
38. Global Fashion Agenda (2022) Fashion on climate: driving the industry towards net-zero emissions. https://www.globalfashionagenda.com
39. Circle Economy (2021) Circular textiles report: transitioning towards sustainable textile systems in Europe. https://www.circle-economy.com/resources
40. Soil Health Institute (2022) The regenerative cotton fund. https://soilhealthinstitute.org/regenerative-cotton-fund/
41. Fashion Revolution (2021) Fashion transparency index 2021. https://www.fashionrevolution.org/about/transparency/
42. Benyus JM (2002) Biomimicry: innovation inspired by nature. Harper Perennial
43. Venkatesan S (2020) Crafting traditions: textile heritage, knowledge, and sustainability. South Asia: J South Asian Stud 43(1):12–27
44. Burgess R (2019) Fibershed: growing a movement of farmers, fashion activists, and makers for a new textile economy. Chelsea Green Publishing
45. Mehta R (2019) Handloom as a site of resistance and renewal in India. TextE: J Cloth Cult 17(4):392–410
46. Bhambra GK (2022) Colonialism and modernity: decolonizing heritage fashion. Cult Sociol 16(2):123–140
47. Mang P, Reed B (2012) Designing from place: a regenerative framework and methodology. Build Res Inf 40(1):23–38
48. EPRS (European Parliamentary Research Service) (2022) Environmental impact of the textile and clothing industry: what consumers need to know. https://www.europarl.europa.eu
49. Ellen MacArthur Foundation (2017) A new textiles economy: redesigning fashion's future. https://ellenmacarthurfoundation.org
50. Ellen MacArthur Foundation (2021) Circular design for fashion. https://ellenmacarthurfoundation.org
51. European Commission (2022) EU strategy for sustainable and circular textiles. https://ec.europa.eu
52. https://www.researchandmarkets.com/reports/5687636/sustainable-fabrics-market-size-share-trends#:~:text=The%20Global%20Sustainable%20Fabrics%20Market,industry%20standards%20and%20product%20innovation

Advancing Sustainability in Textiles: Natural Fibers, Bio-Based Materials, and Recycling Innovations

Neeraj Kumari, Radha, and Sunil Puri

Abstract Nowadays, the textile industry is increasingly shifting towards sustainability to minimize its environmental footprint. This transformation focuses on adoption of eco-friendly materials such as natural fibers, bio-based textiles, and innovations in recycling technologies. Natural fibers like organic cotton, hemp, flax, jute, bamboo and others are gaining attention for their renewable properties and lower environmental impact compared to conventional textiles like polyester, non-organic cotton, rayon and other semi-synthetics. These fibers, however, come with challenges related to water usage and land requirements. Additionally, the emergence of bio-based and biodegradable textiles, derived from sources like algae, mushrooms, microbial cellulose, and fungal mycelium, presents a promising alternative to traditional fibers. These materials are valued for their environmental benefits, yet obstacles such as scalability, performance, and market acceptance need to be addressed for their broader application. Further, recycling plays a pivotal role in circular economy of textiles, contributing to reduction of waste and resource consumption. The recycling of materials such as cotton, polyester, and blended fabrics has seen advancements in technology, although challenges remain in optimizing the efficiency and effectiveness of these processes. The shift towards recycled and upcycled textiles is crucial for reducing the textile industry's overall environmental impact and establishing a more sustainable production cycle. Through this exploration, a comprehensive understanding of current sustainable practices in textile production is provided in this chapter, highlighting the opportunities, limitations, and future potential of these materials and technologies in shaping the future of the textile industry.

N. Kumari · Radha (✉) · S. Puri
School of Biological and Environmental Sciences, Shoolini University of Biotechnology and Management Sciences, Solan 173229, Himachal Pradesh, India
e-mail: radhuchauhan7002@gmail.com

N. Kumari
e-mail: neeruguleria1532001@gmail.com

S. Puri
e-mail: sunilpuri@shooliniuniversity.com

S. S. Muthu (ed.), *Resource Mixes in Textiles*, Sustainable Textiles: Production, Processing, Manufacturing & Chemistry, https://doi.org/10.1007/978-3-032-10921-7_5

Keywords Sustainable textiles · Natural fibers · Recycled materials · Circular economy · Bio-based textiles

1 Introduction

In the last few decades, textile industry is undergoing under significant change as environmental and sustainability concerns reshaping manufacturing practices, consumer behavior, and regulatory frameworks [67, 82]. It is well known that textile industry contributes ecological footprint-including high water and energy demand, chemical discharge and significant amount of landfill waste [53]. According to Mushtaq [107], textile industry cause significant pollution to air, water and soil, with global fiber consumption of 53 million tonnes, annually. Further, it accounts for 10% of global greenhouse gas emission and 20% of water pollution, while only 1% of textile are recycled into new fabric [95, 140]. Additionally, the need for sustainable transformation is due to the estimated increase of 63% in textile consumption by 2030, which is mainly due to fast-fashion trend and disposable clothing trends [49]. This imposes an urgent need to move forward toward circular and sustainable largely driven by the expansion of fast fashion and disposable clothing trends [139], there is an urgent need to explore sustainable alternatives in textile production.

To mitigate these environmental impacts, sustainable textile practices are being prioritized, including adoption of eco-friendly materials, recycling technologies, and innovative design approaches, alongside ethical labor practices and resource-efficient production [125, 145]. One key strategy involves adopting biodegradable and renewable fibers, such as organic cotton, bamboo, hemp, and jute, which require significantly fewer pesticides and less water than conventional fibers [81]. Additionally, advancements in textile recycling—including mechanical and chemical recycling—help reduce waste and extend the lifecycle of textile materials [45, 177]. Circular fashion is also gaining traction, with consumers increasingly favoring second-hand, recycled, and upcycled textile products as sustainable alternatives [60, 145].

Regulatory frameworks are playing a crucial role in shaping the future of sustainable textiles [149]. Governments, especially European Union (EU), are responding with aggressive policy reforms. The Ecodesign for Sustainable Products Regulation, proposed by the EU in March 2022, mandates sustainable design principles, to improve textile sustainability [22, 63]. Additionally, the Empowering Consumers for the Green Transition Directive demands product transparency by mandating product information on durability, repairability, and environmental impact while banning misleading sustainability claims [63]. These measures, along with the Circular Textiles Policy Review adopted in 2023, underscore the urgent need for systemic change in the textile industry [20].

In response to these challenges, industries and startups are developing innovative solutions, particularly in replacing synthetic fibers, which contribute to pollution and a high carbon footprint. According to report (2025–2030), the global natural fiber market which valued at approx. 69.23 billion USD in 2024, is likely to raise at compound annual growth of 5.4% in 2025 to 2030, driven by increasing use of environmental sustainable materials. Leading companies in this sector include Vardhman Textiles Ltd., ANANDHI TEXSTYLES, Grasim Industries Ltd., Bcomp, The Natural Fibre Company, Procotex, Bast Fibre Technologies Inc., FlexForm Technologies, Barnhardt Natural Fibers and Lenzing AG. These firms focus on developing sustainable and high-performance natural fiber composites [112].

Sustainable textiles encompass a broad spectrum of innovations, including natural fiber adoption, bio-based and biodegradable materials, and advanced recycling technologies. Primary plant-based fibers, such as cotton, jute, hemp, and kenaf, offer sustainable alternatives, while secondary fibers-such as banana, coir, and oil palm fibers—are derived as byproducts of agricultural production [25]. These fibers, composed of cellulose, hemicellulose, lignin, wax, and pectin, provide a viable alternative to conventional textile materials. Historically, straw fibers have been utilized in composite manufacturing, demonstrating their potential in enhancing material performance [129]. Given their low cost, high strength-to-weight ratio, renewability, and minimal energy requirements for processing, plant-based fibers are emerging as pivotal materials in the sustainable textile revolution [162]. Therefore, the current chapter explores the evolving landscapes of sustainable textiles-from natural fiber innovations and bio-based alternatives to advanced recycling systems. This chapter focuses on the objective to provide comprehensive overview of how these developments are reshaping the textile industry in alignment with sustainability goals to protect the environment. The chapter discussed the challenges in textile industry, highlighting the use of bio-based and natural fibers and also upcycling and recycling technological advancements. The chapter also reviews the environmental impacts through life-cycle assessment, policies and future recommendations for advancing sustainability in textile industries.

2 Sustainable Natural Fibers: The Future of Green Textiles

Natural fibers are at the forefront of the transformation towards sustainable textile production, as name suggest natural is derived from natural sources like plants. The fibers obtained from renewable biological sources like plants and animals. These fibers possess characteristic properties that make them unique in achieving sustainability goals. Natural fibers are biodegradable and possess less environmental impact as compared to synthetic fibers such as polyester and nylon. Moreover, utilization of natural fibers has potential in textile industry due to eco-friendly nature, will reduce the reliance on fossil fuels, results in lower carbon footprint and alignment with principles of circular economy [36]. Within plant-based fibers, materials like organic cotton, hemp, jute, flax and bamboo are most frequently investigated and utilized in textile industry. Chemically, these fibers are composed of cellulose, hemicellulose, and lignin- kind of polymers that contribute to the mechanical strength and biodegradability of fibers. Other sources of plant origin fibers includes kapok, sisal, banana, pineapple, kenaf, ramie, etc. used to manufacture different products such as hats, bags, shoes, and caps etc. [70].

2.1 Cotton

Cotton remains the most widely used natural fiber globally. It is preferred over other fibers by the consumers due to its characteristics like softness, breathability, texture, water absorption and temperature-regulating properties. Additionally, it is easy fiber that is hypoallergenic and friction free [70]. Among the natural fibers, cotton is appreciated for its biodegradable nature, which minimizes environmental harm. However, conventional cotton cultivation has significant environmental drawbacks. It is a resource-intensive crop that requires vast amounts of water and land and is considered one of the most pesticide-intensive crops globally. Additionally, pre-treatment process of conventional cotton utilizes large amount of chemicals and energy sources [24, 55, 171] (Fig. 1). For example, according to the report of World Wildlife Fund, to produce single T-shirt made of cotton requires 2,700 L of water. Additionally, farming of cotton accounts for 24% of insecticide use and 11% of pesticide sales of world and about 1 kg of cotton can produce upto 3.6 kg of CO_2 [38]. The residues of these agrochemicals may persist on fabric even after washing, raising concerns about exposure leading to potential health and environmental risks. Additionally, cotton processing involves chemical-intensive stages such as scouring, bleaching, and mercerization, which use caustic soda, formaldehyde, chlorine-based bleaching agents, acetic acid, and aqueous sodium hydroxide that contributes to water pollution [56, 153, 168]. In response to these concerns, the production of organic cotton has emerged as a sustainable alternative. It is cultivated without synthetic inputs, reduces aquatic eutrophication, greenhouse gas emission and toxicity impacts, offering improved handleability and dimensional stability while reducing the ecological footprint of cotton production [35].

2.2 Jute

Jute fiber, a plant-based multicellular material. It is a bast fiber distinguished by its cross markings and nodes in longitudinal section and while in cross-section has polygonal shapes. It is cultivated in countries like India, Myanmar, China and Bangladesh and it requires about 4 months to reach a height of about 15–20 cm [70]. Jute is highly hygroscopic (12–14% moisture) and varies in color from gray or yellow or brown. Additionally, it possess good insulating properties making it suitable for diverse applications. As a renewable resource, jute is harvested annually and generates substantial biomass while requiring minimal synthetic pesticides or fertilizers. Its natural biodegradability and recyclability make it an excellent choice for sustainable products. Beyond its ecological benefits, jute exhibits high tensile strength,

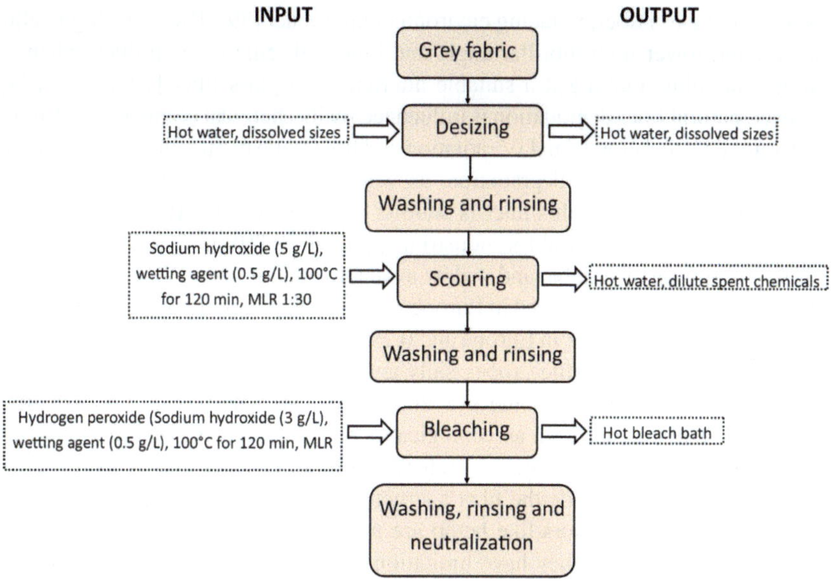

Fig. 1 Pre-treatment process of conventional cotton, involving the inputs and outputs viz. desizing, scouring and bleaching (MLR: material-to-liquor ratio) (adapted from [55])

luster, long staple length, and low extensibility, along with notable resistance to heat and fire [133]. It is widely used in the production of hessian, burlap and gunny cloth, serves as raw materials for jewelry and handbags. However, despite these advantages, the processing of raw jute involves chemicals like sulfuric acid, sodium hydroxide, soda ash and hydrogen peroxide, for cleaning and bleaching, which can have adverse environmental effects [6].

2.3 *Hemp*

Hemp, derived from *Cannabis sativa*, is gaining renewed interest for its strength, stiffness, and fast-growing properties. Hemp fiber has been utilized by humans for thousands of years due to its versatility and sustainability. The plant's fiber-rich stalks are particularly valuable, as they are used in producing paper, rope, and textiles, among other applications. With the increasing demand for eco-friendly and durable materials, hemp fibers have gained popularity as reinforcement in composite materials [36].

These fibers are found abundantly in stem of plant, well known for their exceptional strength and stiffness, making them ideal for composite reinforcement [148]. Compared to traditional glass fiber-reinforced composites, hemp fiber mats contain less resin and glass fiber, reducing environmental impact [89]. Due to its high cellulose amount, lower microfibrillar angle and long cell length, hemp fiber enhances Young's modulus, making it a suitable alternative to glass fiber [124, 159, 175]. However, natural fiber degradation is influenced by its chemical composition. Lignin accelerates deterioration from UV radiation and fire, while hemicelluloses contribute to biological and thermal degradation, as well as moisture absorption [98, 105]. Despite these challenges, this fiber is sturdier and tougher than E-glass, and ramie making it a potential material for industrial applications [101].

Hemp is also an affordable and widely available natural fiber. While some strains of *C. sativa* are legally cultivated in Europe and Canada, regulations limit tetrahydrocannabinol content to 0.3% in Europe and 0.2% in Canada [120]. Historically, hemp fibers have been used in textiles, ropes, sails, and even paper-early Bibles were printed on hemp paper [131]. Finally, after use, these fibers can be burned as a eco-friendly and renewable resource. From an ecological perspective, reusing and mechanically recycling hemp fiber composites is preferable to thermal processing, as recycling through heating can degrade the fiber's structural integrity. Life stage studies indicate that while natural fibers like hemp are more sustainable than glass fibers for lightweight applications, they have limitations in recyclability [121]. Nonetheless, hemp fiber remains an environmentally friendly alternative with significant industrial potential.

2.4 Bamboo

Bamboo is another promising eco-friendly and smooth-textured material that seamlessly integrates sustainability into its applications. Its fibers are widely utilized in textiles, paper production, and construction materials due to its strength and biodegradable nature. Thus, the increasing demand for bamboo fiber aims to reduce or eliminate the dependence on non-renewable glass fibers [87, 94]. Beyond its mechanical benefits, bamboo fiber offers antibacterial, bacteriostatic, and ultraviolet-resistant properties, eliminating the need for additional chemical treatments in finished products [172]. It also exhibits remarkable mechanical characteristics due to its low cellulose content, microfibrillar angle, and high volume resistivity, making it different from other fibers. These properties make bamboo composites suitable for use as dielectric materials [84]. Several bamboo varieties, including *Phyllostachys makinoi, Bambusa stenostachya, P. pubescens* and *Dendrocalamus lactiferous*, are used as reinforcements in polypropylene mixtures/composites. Studies indicate that Makino bamboo significantly influences tensile and modulus properties due to its high crystallinity and lignin content compared to other bamboo types [173]. Recent advances in thermoset polymer material composites using bamboo fibers have shown industrial applications in green building materials and technical textiles as adhesives, coatings, and insulation materials [83, 128]. Research on bamboo-based composites with variable polyurethane concentrations (10–25 wt%) shows that increasing matrix concentration enhances mechanical performance. Experiments involved soaking bamboo fibers in phenol–formaldehyde solutions for 4–8 min, followed by drying and hot pressing at 150 °C under pressures of 3.5–7 MPa. Mechanical tests, including compression and bending, confirmed improved strength with higher matrix content. The mechanical performance of thermoset polymer composites incorporating bamboo fibers is influenced by fiber aspect ratio, adhesion, content and distribution, alike to thermoplastic matrices [132].

2.5 *Other Fibers*

Sisal, derived from *Agave sisalana*, is a durable and adaptable natural fiber widely used in sustainable textiles. As a lignocellulosic fiber, sisal is renewable, biodegradable, and serves as an excellent reinforcing material in thermoplastic and thermosetting polymers [102]. It is used for ropes, mats and composite reinforcements. Additionally, sisal fiber has been integrated into cement and geopolymer composites for structural reinforcement, offering an environmentally friendly alternative to synthetic fibers [5]. Studies on geopolymeric composites reinforced with sisal have demonstrated strong flexural strength, microstructural integrity and durability, even after accelerated wet-dry aging cycles [16]. Flax is another promising sustainable fiber used in textiles and composites. Its hydrophilic nature results from its chemical composition, which includes cellulose, hemicellulose, and lignin [106]. Flax fibers are two-timing as strong as cotton and is considered one of the best plant-based alternatives to synthetic fibers [47, 169]. Due to its renewable and biodegradable properties, flax fiber is increasingly used in automotive applications, reducing energy consumption by 83% and costs by 40% compared to glass fiber-based components [46]. These fibers, along with bamboo, jute, and hemp, offer sustainable solutions across multiple industries, reinforcing the shift towards eco-friendly and cost-effective materials. Table 1 provides the annual production rate of natural fibers, globally.

Table 1 Annual production of natural fibers

Natural fibers	Annual production (Ton)	References
Jute	2,300,000	[9, 143]
Bamboo	10,000,000	
Hemp	214,000	
Sisal	378,000	
Others-wood: softwood kraft pulp: wheat	1,750,000,000: 26,000,000: 720,000,000	
Cotton	18,450,000	

3 Bio-based Textiles: The Next Generation of Sustainable Materials

As the textile industry moves towards reducing dependence on fossil fuels or non-renewable sources, bio-based textiles have emerged as an evolutionary material to achieve sustainable goals. The biobased fibers extracted from renewable biological sources like moss, bacterial cellulose, holds a significant potential as a raw material in textile. These fibers form part of what is now recognized as third generation of textiles-engineered for renewability, biodegradability and functional performance [161].

Bacterial cellulose (BC) is a nanostructured bio-material produced by certain strains of bacteria. It owns a reticulate 3D weaves network of cellulose, along with significant properties like mechanical, suspension stability and water holding capability. It has gained attention in both industrial and academic field, currently explored as a source for plant-less rayon and fabric. Plant-based renewed fibers such as cuprammonuium rayon, derived from wood pulps and cotton, while it is biodegradable utilizes large amount of energy and chemicals during processing [75, 137, 178]. An Australian company, "Nanollose Ltd." is exploring BC to convert it into eco-friendly fibers for industrial applications. They have successfully converted BC into viscose-rayon fibers, providing alternative in textile having numerous advantages as compared to other fibers [111]. Nanollose has transformed BC into Nullarbor ™ using their own advanced developed technique which is further made into cloth and fabric [111].

Alage-based textiles represents another advancement in sustainable textile approach. For instance, seaweed-derived alginate (*Saccharina latissima*) shows promising results in textile industry as an alternative to other materials in market. The derived product can be used in partially elastic fabric materials such as jeans, skirts, sportswear and waistbands [14]. The consideration of using algae as biofabric material involves certain benefits such as high quality material that uses less chemicals during processing, degrade faster than cotton, does not require pesticides and a lot of space to produce [13]. Eponymous label has collaborated with Charlotte Mccurdy (a scientist well known for producing algae plastic from organic sources in ocean) to create a sequin dress made of algae, promoting alternatives in textile industry to address issues related to environment [54, 115]. Moreover, numerous studies has proven the potential of bacterial cellulose as a sustainable raw material in textile industry. The application of bacterial cellulose ranges from basic clothing to wearable electronics. The use of bacterial cellulose aligns with Sustainable goals 6, 7, 12, 13 and 15 [114]. Suzanne Lee, a British designer has developed the idea of using fermented yeast, bacteria and sweetened green tea for the production of fermented clothes. However, there are some limitations to this idea as the developed fabric is not water resistant, absorbs 100 times its weight in liquid and not ideal for everyday wear [156]. Therefore, there are further need to explore/investigate the processes, material and others to utilize these-biobased products for everyday use and other applications.

Additionally, Kalayci et al. [77] reported that *Tillandsia usneoides* (a Spanish moss) holds potential in textile industry. Its characteristic properties are responsible for comfort, durability, shock absorption, breathability and thermal insulation. This moss also promises environmentally friendly production, as it grows self, absence of chemicals or toxic substances in production, and biodegradable properties [77]. Fungal-mycelium-another bio-based material gaining attention in textile innovation. MycoWorks, a biotechnology company is working on producing high quality natural materials (leather) to match the performance of finest animal leathers-using Fine Mycelium™ [110]. Additionally, their applications extend beyond textiles into packaging, furniture and others. Table 2 provides an list of advantages of using natural fibers in production of textile products.

However, integration of synthetic biology with AI-based material simulation and closed loop systems holds significant potential in advancement in textile industry. Collaboration of fashion brands with biotech companies may achieve environment sustainability goals by adopting and promoting natural fibers and bio-based materials and also helps in terms of cost, performance and fashion trends.

Table 2 Overview of advantages and limitations of natural fibers if used in textile industry

Advantages	Limitations	References
Renewable resources	Non-homogenous structure of fibers	[9, 50, 65, 147]
Require lesser cost for production	Structural instability due to water absorption	
Possess significant mechanical properties	Lower resistance to heat and water	
Results in composites of lower density	Susceptible to attack and rotting because of higher water absorption ability	
Utilizes less energy during manufacturing	Non-significant adhesion and incompatibility with polymer matric	
Eco-friendly and biodegradable material	Aging and degradation	
Does not pose any harm to human lives	Still require chemicals during processing	
Lower Greenhouse gas emission	Variability in emissions reduction depending on cultivation, processing, and transport logistics	
Uses less chemicals as compared to synthetic fibers	Require chemical pretreatments (e.g., for retting, bleaching, dyeing), which can reduce eco-benefits if not managed sustainably	

4 Recycling and Upcycling of Textiles

In the past few years, the industries mainly textile is currently undergoing smart waste management practice by adopting sustainable practices such as recycling and upcycling that closely aligns with principles of circular economy [29, 170]. This transformation aims to reduce amount of waste generated and improving the use of resources across all stages of production. In contrast to traditional model (take-make-dispose), the circular systems focuses to extends product life, maximize resource utility and reduce waste. Within this context, recycling and upcycling addressees two major concerns related with textile industry: raw material substitution and post-consumer waste management [29, 151, 170].

Globally, 92 million tonnes of textile waste is generated every year. With production doubled from 2000 to 2015 and duration of garment decreased by 36%, only 8% textile fibers in 2023 are made from recycled sources [163]. In view of this, many industries are focusing on durability, reusability, repairability, and recyclability of products, contributing to reduction of environmental footprint from textile sector [36, 130]. For example, Sodra, an international forest industry group, developed Once-More company that recycles blended fabrics, more likely polycotton blends. The cotton is recovered from polyester blend and reused by combining it with renewable wood cellulose, converting waste into high-quality material. Additionally, the remaining polyester is incinerated for energy production, which powers OnceMore process enabling circular flows in textile and fashion industry [126].

However, due to current technological limitations in textile processing, there are limitations to the introduction of recycled material in new products. For example, the proportion of recycled cotton used in new yarns is generally restricted to about 30%. This constraint remains a significant barrier to producing high-quality textiles solely from recycled cotton [34]. Similar limitations are also observed in the recycling of synthetic fibers, such as polyester. Additionally, this method reduces the reduces dependence on fossil-fuel-derived virgin polyester and lowers energy consumption as compared to conventional production processes [85, 118, 142].

A further complication in textile recycling arises from the increasing use of blended fabrics, such as cotton-polyester (polycotton) combinations. These blends, while offering enhanced functional properties, are difficult to recycle due to complex interweaving of different fiber types [95]. Mechanical recycling techniques does not typically help in separating the fibers effectively, whereas chemical recycling can be applied to recover targeted fiber from blends by depolymerizing or dissolving a particular polymer [27, 95]. Recent research into solvent-based and enzymatic separation methods has shown potential, but the high expense cost and complex working operational complexities limit their widespread application [6, 95].

Upcycling involves transforming waste textiles into products of equal or higher value than the original material. Unlike recycling, upcycling preserves the integrity of material while reimagining its use. Comparative analyses suggest that upcycling has a lower environmental impact and supports a more circular approach by encouraging design innovation, consumer engagement, and local-level production [73, 176].

Both recycling and upcycling contribute significantly to circular economy by conserving resources, minimizing waste, and prolonging lifespan of textiles [139]. Recycling is important for material recovery but often requires high energy consumption and may lead to downcycled products with reduced functionality [27]. In contrast, upcycling is generally less resource-intensive and offers advantage of preserving or enhancing the value of material [73, 157]. However, several challenges limits the adoption of recycling and upcycling, including the complexity of blended fibers, low consumer awareness, economic limitations, and inadequate infrastructure for large scale sustainable textile waste management [27, 95]. Therefore, integrating both practices within the textile industry is not only feasible but also essential for reducing environmental footprint, conserving resources, and promoting long-term sustainability.

4.1 Recycling Processes for Textiles

Recycling is typically categorized into mechanical, chemical and biological methods, each having advantages and limitations.

4.1.1 Mechanical Recycling

It is one of the most frequently used method in waste management in textile industry due to its low cost and scalability. However, mechanical process results in materials with lower fiber quality, limiting their applications for their reuse [140]. This process is particularly advanced for cotton, allow semi-closed loop system, involving mixing or blending with virgin cotton to form yarns, unlikely to synthetic fiber involving open-loop outcomes [37, 40].

Mechanical Recycling at Fiber Level

Generally, fabrics or textile products are shredded or cut after removing non-textile components (buttons, zippers), requiring manual labour [1]. Then the shredded fabric is processed into slivers that made into yarns [140]. Shredding weakens length of fiber, therefore blending it with virgin fiber is usually necessary to reuse, meeting the quality standards [91].

Thermomechanical Method

Also known as thermal recycling or mechanical polymer, involves melting of fibers for remolding or re-spinning [116]. It is mainly employed for synthetic fibers like nylon, elastance, thermoplastic polyurethane [86], but it is limited due to brittleness unless transformed into solid products [117]. However, technologically this process is limited due to high purification cost with exception such as Teijin's polyester recycling initiative in Japan [96].

Natural Fiber Recycling

This process involves extraction of fibers from agri-waste and converting them into textiles through mechanical methods. For instance, Pina fabric (or Pinatex) processed from pineapple leaves [8, 74], Agraloop BioFibre uses rice husk, banana stalks and flax straw to create biofibers through hybrid process involving wet spinning [3]. Other sustainable approaches include Spinnova's fiber suspension technique, processing crop residues like straw into textiles [138, 141] and cellulose nanofiber from sugar beet and potato have also been explored for their potential in textile industry [62].

Environmental Impact of Mechanical Recycling

Despite the growth of the textile industry, the environmental data is limited, therefore energy consumption can be regarded as preliminary indicator. It was reported that virgin cotton utilizes about 50 MJ/kg while recycling uses 5–20% of this, which is significantly lower. This highlights that mechanical recycling is energy efficient, especially when high intensity steps like dye removal or chemical treatment are not involved [108, 144].

4.2 *Chemical Recycling of Textiles*

This method involves the use of chemicals for breaking down textile polymers into monomers or preserving the polymer structure for regeneration. It enables recovery of high-quality fibers as compared to mechanical method. It is mostly used for cellulose based and synthetic fibers. However, chemical recycling is still emerging commercially due to relatively high cost, technical challenges and chemical toxicity [139].

4.2.1 Monomer Recycling

It breaks down the polymer chains into individual monomers, which are then repolymerized into new quality fibers. Although, chemical recycling of cellulose has been explored, it has yet to be used in commercial textiles. Some synthetic fibers like polyethylene terephthalate, nylon 6 has been recycling using the chemical recycling method [41, 44, 51].

4.2.2 Polymer Recycling

In this process, fibers are physically broken down (shredded), dissolved in chemical solvents. This help in preserving polymer structure, allowing regeneration into new fibers [150]. Lenxing's lyocell process (using N-methylmorpholine-N-oxide) is considered environmental friendly due to lower toxicity and recyclable solvents [72]. This method involves the used of pre- and post-consumption cotton waste [58, 158]. Ionic liquids, a new glass of green solvents, are also used to solubilize cellulose from textile [12, 97]. Technologies like Ioncell-F aims to commercialize this route, combining high efficiency with low environmental impact. Commercial ventures like Circulose and Nucycle are focusing in upscaling cellulose-based chemical recycling [30, 42]. Cellulose from orange fiber in Italy and sugarcane residues have also been done using the similar approach [31].

4.2.3 Separation of Fiber Blends

Separation of fiber blends (e.g., cotton-polyester) is one of the key challenge during recycling. Chemical recycling techniques focuses on single-material input, therefore the separation of blends is essential. Technologies like OnceMore (Sodra) separates polyester from cotton-polyester blend, enabling recovery of cellulose [119]. Other practices like using sodium hydroxide to depolymerize the polyester fraction or using ionic liquid to dissolve cellulose component has also been used in textile recycling [57, 174]. Additionally, advance separation methods can achieve upto 98% recovery rate -with cellulose fraction repurposed for ethanol production [48]. Some ventures like Worn Again, H&M (in association with Hong Kong Research Institute for Textile and Apparel) are working to commercialize blend-separation solutions [41].

4.2.4 Environmental Impact of Chemical Recycling

Although, chemical recycling ensures high recovery rate, even from the textile blends, it raises environmental concerns. It involves use of toxic chemicals and higher electricity consumption. The chemical recycling of polyester consumes about 12 MJ/kg, higher than mechanical recycling [144]. However, scaling up these technologies require regulatory support, investment and optimization schemes for both cost and environment burdens [11].

4.3 Bio-Based Processes for Textile Recycling

This involves use of biological agents such as microbes, enzymes, or fermentation technique to convert textile materials. However, these processes are promising due to their low energy demand, renewable sources and potential to replace harmful conventional chemical methods. The bio-based recycling process involves, enzymatic depolymerization, decomposition, and fermentation method.

Biological decomposition involves anaerobic digestion and composting, where organic textile waste is converted into simpler compounds. Composting includes aerobic microorganisms that convert textile into waster, heat and carbon dioxide, producing soil-quality enhancing compost. The process of composting has been proposed for plant-based or regenerated cellulose fibers [19, 41]. In case of anerobic digestion, microbes convert cellulose-rich materials into carbon- dioxide and methane [76, 136].

While enzymatic depolymerization process involves the use of enzymes such as cellulolytic enzymes (e.g. cellulases) to breakdown cellulose in textile into glucose monomers. The enzymatic processes require prior chemical pretreatment like alkali treatment (sodium hydroxide) or use of ionic liquids or green solvents, to reduce crystallinity and improve enzyme accessibility [17, 97, 152]. Recent advancements includes enzymatic hydrolysis combined with low-temperature alkali pretreatment to convert cotton-polyester blends into recoverable components via low-temperature alkali pretreatment [93, 154]. Cellulase enzymatic blends from companies like DuPont and Novozymes are already used in textile biorefinery research, allowing the selective hydrolysis of cellulose. This approach leads to the improvement in separation in blends without manual sorting [166].

Fermentation process use microbial metabolism to convert organic material into useful products such as ethanol, enzymes or organic acids. Fermentation is also of three types: (i). solid state fermentation takes place in minimal free water and sustainable for producing cellulase e.g. *Trichoderma reesei* to ferment textile and recover polyester [64], (ii). submerged fermentation-converts textile derived sugars into ethanol [69], succinic acid [93], and other biochemicals [48, where one or more engineered microbes handle both hydrolysis and fermentation steps [133], (iii). Biopolymer manufacturing-involves microorganisms like *Komagataeibacter xylinus* to produce bacterial cellulose and process into fibers. Thus, the bacterial cellulose can be used as an alternative for textile or leather substitutes [100, 134]. Nanollose company has demonstrated the bacterial cellulose as input for spinning into regenerated cellulose fibers [111].

4.3.1 Environmental Impact of Bio-based Processes

Biobased recycling process favors low energy demand, mild processing conditions and use of renewable energy sources. However, the sustainability of these processes depends on different factors like chemical usage during pretreatment, energy demand for mixing or cooling and enzyme production cost [66]. However, from above discussion it can be concluded that the integration of mechanical, chemical and bio-based recycling processes represents a critical step in advancing sustainability within textile sector. While mechanical recycling offers low energy demand, scalability, its limitation in fiber quality call for complementary approaches. Chemical recycling, with its potential for high-value fiber regeneration, is evolving rapidly but requires solutions for high cost, solvent toxicity and blend separation. Meanwhile, bio-based innovations-including hydrolysis, fermentation and microbial polymer synthesis-offering low impact alternative aligned with circular economy principles. Collectively, these innovation underscores the need for development in the field of infrastructure, cost, and supportive policies to overcome current technological barriers. However, advancing these recycling strategies will be pivotal in reducing dependency on virgin resources, minimizing environmental impacts, and fostering a regenerative textile system that aligns with the principles of circularity and long-term sustainability.

5 Environmental Benefits and Limitations of Sustainable Fibers

As aforementioned earlier, textile industries are adopting more sustainable approaches. The introduction of natural fiber in textile industry has proved as a potent alternative to synthetic fiber which is related with environmental footprint. However, with growing awareness both fiber producers and fabric manufacturers are also exploring other alternatives to meet the requirement of sustainable practices. Additionally, impact varies significantly based on agriculture practices, geographical location, processing method and life-cycle stages [113].

A key factor involved in textile production is the selection of raw material which significantly determines the overall environment footprint of developed product [113]. This decision shapes land impact, water usage, pesticide reliance, energy consumption and eco-friendly nature. For example, organic cotton results in decreased aquatic eutrophication and GWP impact over conventional cotton [146, 147]. Also, it does not cause carcinogenic risk to human livelihoods [65]. Jute has an advantage over cotton, requires much less water and results in lower impacts like energy demand, water resource depletion, human toxicity and acidification [90]. Flax utilizes lower inputs but results in higher yields, minimum GHG emission and energy and water usage in comparison to cotton [18]. Since certifications of both jute and flax is limited, it can be assumed that fertilizers and insecticides used during their cultivation contribute to environmental impact [50]. Comparatively to all natural fibers, polyester consumes large amount of electricity as reported by Khabbaz [79] than flax, conventional cotton, wool, and polypropylene. It also contributes to terrestrial ecotoxicity, climate change as reported by Moazzem et al. [104].

Other fibers like banana, hemp and bamboo etc. are also under investigation for their use in textile industry as it offers promising alternatives as compared to cotton due to their lower demand for water and chemicals [7, 80, 135]. Additionally, innovations in fiber extraction from waste agriculture sources like coffee grounds, orange peel, banana pseudostem, and pineapple leaves represents a solution aligned well with the concept of circular economy, reducing overall resource demand and valorizing agri-waste [4, 164].

Recently, life cycle assessment (LCA) studies further highlight the concerns in textile industry [50] (Table 3). It has emerged as a standardized tool to compare fibers based on their environmental performance across life stages-viz., from raw material extraction and fiber processing to textile manufacturing, distribution, customer use and end of life [109]. However, expanding the research on LCA will help the stakeholders in identifying more sustainable fibers and excluding the ones which are less environment friendly (Table 3). It can guide customer towards more informed decisions in purchasing and responsibly disposing them when no longer in use [50].

6 Brand Utilizing Natural Fibers

Numerous fashion brands are adopting natural fibers to reduce environmental impact and offer healthier alternatives to synthetic fibers or textiles. These brands are utilizing banana, linen, hemp and organic cotton in their fabrics. MATE, a clothing brand based in Los Angeles produces clothing essentials using organic cotton, linen, Tencel. The brand claims that their products are made with non-toxic materials, using organic material and dyes and also recycle old garment to new. The MATE label is certified as Climate Neutral, GOTS certified and a B corporation [103]. Eileen Fisher, a New York based brand using organic pima cotton and also merino wool. It focuses on regenerative farming practices, repurposes and upcycles excess textiles. The brand use solar-powered warehouses for production [39, 160]. Farm to closet initiative of Christy Dawn provides eco-friendly clothing with a focus on regenerative cotton dresses. The brand claims that each dress is made from cotton grown using traditional, organic methods that sequester carbon. The brand uses organic and deadstock fabrics, highlighting their commitment to sustainable and earth friendly fashion [32, 160].

Colorful Standard, Organic Basics, Harvest & Mill, KOTN, Jungmaven, Boody, The Standard Stitch, Yes Friends, Outerknown, tentree, Pact, No Nasties, vanaras, A Big Indian Story, The Summer House, Ambiya, B Label etc. are some of the international and national brands that are promoting the use of sustainable fibers in textiles [10, 160].

7 Future Perspectives and Innovations

The advancements of textile industry lie at the convergence of technological innovations, systemic reforms and consumer-driven transformations. As the demand for fabrics continues to rise, emission from textile manufacturing is projected to skyrocket by 60% by 2030 [99]. The environmental toll for its production and renewability [155]. To mitigate the negative environmental impact, innovators and companies/brands are targeting every stage of textile life cycle, including eco-friendly

Table 3 Comparative analysis of natural vs synthetic fibers environmental impact in agriculture production processes (*Source* [50])

impact category	Flax	Conventional cotton	Polyester	Jute	Organic cotton
Fine particulate matter formation (kg PM 2.5 eq)	0.00015	0.00068	0.00232	0.00014	4.69950×10^{-5}
Water consumption (m^3)	0.0624	0.36277	0.05444	0.05661	0.00668
Fossil resource scarcity (kg oil eq)	0.02071	0.21201	1.91054	0.00943	0.01708
Terrestrial ecotoxicity (kg 1,4-DCB)	0.20165	2.57065	6.28966	0.10487	0.20721
Freshwater ecotoxicity (kg 1,4-DCB)	0.00743	0.17624	0.12778	0.00491	0.03973
Terrestrial acidification (kg SO_2 eq)	0.00275	0.01514	0.01253	0.00105	0.01048
Freshwater eutrophication (kg P eq)	3.00837×10^{-5}	0.00097	0.00109	0.00013	0.00611
Global warming (kg CO_2 eq)	0.17942	1.35726	5.08799	0.14672	0.46731
Stratospheric ozone depletion (kg CFC11 eq)	1.27404×10^{-6}	7.21148×10^{-6}	4.48697×10^{-5}	4.93978×10^{-7}	8.28099×10^{-6}
Human carcinogenic toxicity (kg 1,4-DCB)	0.00051	0.00028	0.00133	0.00056	0.00019
Ozone formation, Terrestrial ecosystem (kg NOx eq)	0.00071	0.0057	0.01099	0.00031	0.00361
Human non-carcinogenic toxicity (kg 1,4-DCB)	0.00341	1.9838	0.11459	0.00421	0.00892
Ionizing radiation (kBq Co-60 eq)	0.00253	0.02449	0.21618	0.00146	0.02011
Ozone formation, Human health (kg NOx eq)	0.0007	0.00563	0.00994	0.00031	0.00304

(continued)

Table 3 (continued)

impact category	Flax	Conventional cotton	Polyester	Jute	Organic cotton
Marine ecotoxicity (kg 1,4-DCB)	0.00182	0.04104	0.03775	0.0012	0.01228
Marine eutrophication (kg N eq)	0.00068	0.00582	0.00035	0.00034	0.01194
Mineral resource scarcity (kg Cu eq)	0.00068	0.00464	0.00983	0.00027	0.00035

CFC: Chlorofluorocarbon; DCB: Dichlorobenzidine

production and dyeing to recycling and post-consumer recovery (circular system). Concurrently, majority of key stakeholders have adopted sustainable manufacturing practices [28, 52]. Advanced methods such as waterless dyeing, use of bacteria to create vibrant, sustainable dyes, changing the surface chemistry of the fabric via pre-treatment processes the concept, recycling of mixed-material clothing has emerged as a sustainable approach that involves less utilization of resources like water, chemical and energy. According to Greyb Services report, over 1900 patents and 3200 research papers have been published, investigating waterless dyeing method alone, highlighting the scale of investment and interest [155].

After production, one of the primary obstacle after post-consumption is recycling from back to fiber, especially for blended fabrics made into new fabrics that mostly contribute to environmental waste [43]. Consumers increasingly seek products made of natural material such as organic cotton and hemp and are willing to pay premium prices that align with environmental sustainability [155]. It has been inferred from the studies that there is a significant less availability of commercially feasible recycling technologies for low-grade materials, mainly due to less exploration of the developed, large-scale processes for separating fibers in mixed and composite textile locally to small scale industries in abundance. The major downfall of textile recycling at local and regional scale, not only due to cost of establishment and processing but also due to lack of knowledge about the working process. Most of the processes used in recycling are complicated and cannot be understood by the locals due to little knowledge about the technology [43].

To mitigate GHG emission and protecting environment from climate change, textile industry needs to take a step away from restrained decarbonization trajectory towards a more advanced and efficient one [2]. For large scale changes, innovations requires the support through stakeholder collaboration, government policy, infrastructure and investment plans. The solutions involve advancements in technology and models for farmers to support biomaterial production and recycling innovations, including reuse and recycle textiles to form a closed loop system to lower industry's ecological footprint [2, 71].

The textile industry positively results in CO_2 emission. Despite, positive long-run effect of renewable energy consumption, it reduces effect of CO_2 emission from clothing and textile industry [59, 88]. Thus, goal of decarbonization can be achieved via replacing fossil fuels such as coal with more renewable sources [26], utilizing biobased raw materials/feedstocks, sustainable production, low-emission heat sources, maximizing rate of recycling materials and minimizing transportation processes [2] and regeneration of natural materials [69]. Prenzel et al. [127] has reported energy consumption is responsible for climate change 59%, related with carbon fiber results in 67% reduction rate of energy usage by shifting to renewable energy source. Additionally, reducing processing time for production has resulted in 70% reduction in energy utilization [68]. Das [33] observed that using natural fibers rich in lignin emits 22% less CO_2 than polyacrylonitrilie based fiber. Therefore, by achieving net zero electricity usage alone can lead to 50% reduction in global warming potential, highlighting the value of decarbonizing electricity in textiles [122].

Additionally, advancement in 3D modeling, virtual prototyping and AI-driven design tools are revolutionizing how textiles are created. By predicting the precise material required and zero-waste cutting will also reduce material consumption and energy cost. This would also help in reducing the transport costs. The innovation efforts should be placed around waste minimization, carbon footprint measurement and reduction, investment in pollution prevention technology [123].

The success of future innovations depends heavily on the regulatory environment. Government policies for supporting manufacturing units through incentives, subsidies and lower interest rates on loans and also strategies for renewable energy sources should be developed. New transformation and regulation policies including strict supply chain agreements and funding to support better business models [21], also extended manufacturer responsibility legislation to fund collection, filtering, and recycling of post consumption textile needed to be included [21]. Consumer behavior has also begun to shift and demands for products that are fashionable or attractive and functional have increased. In view of this, it is expected that textile industry/ brands which are well known- should play a more prominent part in customer choice [92].

In conclusion, the textile industry needs to shift towards "less but better" through minimum cost value, high quality and constant development to accelerate growth. Marketing enterprises which encourage consumption also need to be upgraded. Consumer behavior towards sustainability should be promoted through eco-friendly use, recycling, rental and repair garment practices, reuse and sharing options [2, 116].

8 Policies and Tools for Environment Sustainability in Textile Industry

Policy frameworks play an indispensable roles in catalyzing the transition of textile industry towards sustainability. Government, intergovernmental organization and other regulatory are increasingly issuing regulations and guidelines to promote resource efficiency, circular economy and transparency across textile value chains. For example, California's Responsible Textile Recovery Act of 2024 encourage the clothing retailers to provide free recycling program from unwanted textile, makes them responsible for collection, repair and recycling, setting an example for legislative action in United States [61, 78]. Additionally, initiatives such as the introduction of digital passports for fabric/textile to enhance transparency in origin, supply chain, allowing better consumer decision during purchase [167]. These combined efforts are helping in forwarding textile industry towards a more sustainable and circular economy, aligning production and consumption with environmental benefits.

Additionally, UN Environment Programme (UNEP) textile initiative addresses strategic leadership enabling them to integrate policy changes towards circularity and sustainability by supporting mapping policies, providing policy blueprints, fact-sheets, developing recommendations and analysis on textile trends. UNEP also works to phase out hazardous chemicals, making sector contributing to industry under Global Framework on Chemicals. It supports work in regions including, Cambodia, Vietnam, Indonesia and Trinidad & Tobago, to reduce consumption and release of chemicals of concerns across whole supply chain and shift towards circular business models. It also addresses concerns of using textiles to differentiate between using textiles and textile waste [165].

Indian government has also taken initiates by forming regulatory frameworks that works for reducing the environmental pollution involving air, soil and water systems. Therefore, major initiatives have been undertaken under various legislations and schemes on pollution abatement. For example, the Central Pollution Control Board works on principals of (i) prevention, control and abatement of water pollution, (ii) prevent, control air pollution in country and also evaluation of common effluent treatment plants and programs for environmental capacity building and awareness [23].

The scheme of Industrial Pollution Abatement through strategies involves three components: adoption of clean technology in small scale industries and environmental statistics and environmental audit. This scheme assists small scale industries for cleaner production practices and reduction in waste generation [23]. Another, the Environmental Impact Assessment is one of the important management tool for incorporating environmental concerns in development projects at the planning stage. Environment clearance is made mandatory since January, 1994 which was again revised in 2006, highlighting the categorization of developmental activities based on their potential impacts [23]. Further, Common Effluent Treatment Plant enables small scale industries to comply with environmental discharge standards. The programme is spread across India and the funds form the Centre are provided through respective State Pollution Control [23].

9 Conclusion

The transformation or shift of the textile industry is no longer an option-but it is today's necessity, shaped by ecological urgency, consumer awareness and regulatory frameworks. This chapter explores how innovations in textile industry like the introduction of natural fibers, bio-based materials, recycling technologies and global policy frameworks are merging to form a new model for textile production and consumption. The natural fibers like organic cotton, bamboo, hemp, flax, jute etc. have merged as an eco-friendly alternatives to synthetic textiles, offering benefits like low water usage, reduce pesticide consumption, biodegradability and life-cycle assessment. Bio-based innovations such as potential of algae, bacterial cellulose and moss are promising in the field of environment sustainability, although limitations like scalability and performance barriers exist.

The recycling and upcycling play important part in circular economy, enabling resource recovery and waste reduction. However, recycling processes struggle with fiber degradation, high cost, chemical toxicity. Thus, bridging these gaps requires technology investment, infrastructure and changes in consumer behavior. Policy frameworks such as European Union's Ecodesign Regulation to California's textile recovery mandates and India's pollution control mechanisms is contributing towards transparency, life cycle responsibility and close-loop systems. Tools such as digital product development, sustainability certifications, extended producer responsibility are empowering both consumers and brands to make informed ethical decisions. Altogether with sustainability and regeneration models in industries working on different areas will support community resilience, foster productive growth and restore degraded ecosystem.

References

1. Abtew MA, Atalie D, Dejene BK (2025) Recycling of cotton textile waste: technological process, applications, and sustainability within a circular economy. J Ind Text 55:15280837251348664. https://doi.org/10.1177/15280837251348663
2. Berg A, Granskog A, Lee L, Magnus K-H (2021) How the fashion industry can urgently act to reduce its greenhouse-gas emissions. McKinsey & Company and Global Fashion Agenda
3. Agraloop (2023) Transforming left-overs from crops into scalable high-value natural fiber fashion products. https://circularsystems.com/agraloop. Accessed 4 Dec 2025
4. Agrawal HS, Praneetha M, Singh A, Rajpoot AK (2023) A comprehensive survey on production of fabric from the organic extract. 2023 International conference on computing, communication, and intelligent systems (ICCCIS). IEEE, Greater Noida, India, pp 6–12
5. Ahmad J, Majdi A, Deifalla AF, Ben Kahla N, El-Shorbagy MA (2022) Concrete reinforced with sisal fibers (SSF): overview of mechanical and physical properties. Crystals 12:952. https://doi.org/10.3390/cryst12070952
6. Al-Mamun M, Saha CK, Mostofa MG, Miah A, Hossain MZ (2017) Identification of suitable varieties for seed production of jute in non-traditional areas of Bangladesh. Bangla J Plant Breed Genet 30:33–38. https://doi.org/10.3329/bjpbg.v30i1.36531
7. Amjad AI (2024) Bamboo fibre: a sustainable solution for textile manufacturing. Adv Bamboo Sci 7:100088. https://doi.org/10.1016/j.bamboo.2024.100088
8. Ananas Anam (2024) Piñatex featured in CNN's 'Going Green.' https://www.ananas-anam.com/pinatex-featured-in-cnns-going-green/. Accessed 4 Dec 2025
9. Antov P, Savov V, Neykov N (2017) Utilization of agricultural waste and wood industry residues in the production of natural fiber-reinforced composite materials. https://doi.org/10.13140/RG.2.2.35619.35365
10. Anuri Phatak (2025) 20 Indian sustainable fashion brands that are weaving a brighter future. In: GoFynd the closet. https://www.gofynd.com/thecloset/indian-sustainable-fashion-brands?srsltid=AfmBOoorBeJ6mbvQ3GyOxgBKGkI8w-WN0UqwHei8AJ91wkS1m5SXBPFM#google_vignette. Accessed 15 Apr 2025
11. Ostlund A, Wedin H, Bolin LB, Berlin J, Jönsson C, Posner S, Sandin G (2015) Textilåtervinning: Tekniska möjligheter och utmaningar. Stockholm, Sweden
12. Asaadi S, Hummel M, Hellsten S, Härkäsalmi T, Ma Y, Michud A, Sixta H (2016) Renewable high-performance fibers from the chemical recycling of cotton waste utilizing an ionic liquid. Chemsuschem 9:3250–3258. https://doi.org/10.1002/cssc.201600680
13. Ayyanar I, SJB G, Ravindran A, Rajendran R, Arulprakasam A, Vellaisamy B (2023) Algae materials for textile industries. In: Algae materials. Elsevier, pp 231–246
14. Badruddin IJ, Silva MP, Tonon T, Gomez LD, Rahatekar SS (2024) Modulating the properties of brown alga alginate-based fibers using natural cross-linkers for sustainable textile and fashion applications. ACS Omega 9:37002–37011. https://doi.org/10.1021/acsomega.4c03037

15. Baloyi RB, Gbadeyan OJ, Sithole B, Chunilall V (2024) Recent advances in recycling technologies for waste textile fabrics: a review. Text Res J 94:508–529. https://doi.org/10.1177/00405175231210239
16. Batista Dos Santos GZ, Passos De Oliveira D, De Almeida Melo Filho J, Marques Da Silva N (2021) Sustainable geopolymer composite reinforced with sisal fiber: durability to wetting and drying cycles. J Build Eng 43:102568. https://doi.org/10.1016/j.jobe.2021.102568
17. Beig B, Riaz M, Raza Naqvi S, Hassan M, Zheng Z, Karimi K, Pugazhendhi A, Atabani AE, Thuy Lan Chi N (2021) Current challenges and innovative developments in pretreatment of lignocellulosic residues for biofuel production: a review. Fuel 287:119670. https://doi.org/10.1016/j.fuel.2020.119670
18. Bhalla K, Kumar T, Rangaswamy J (2018) An integrated rural development model based on comprehensive life-cycle assessment (LCA) of Khadi -Handloom industry in Rural India. Procedia CIRP 69:493–498. https://doi.org/10.1016/j.procir.2017.11.072
19. Biyada S, Merzouki M, Dėmčėnko T, Vasiliauskienė D, Urbonavičius J, Marčiulaitienė E, Vasarevičius S, Benlemlih M (2020) Evolution of microbial composition and enzymatic activities during the composting of textile waste. Appl Sci 10:3758. https://doi.org/10.3390/app10113758
20. Bour A, Budde Christensen T, Hunka AD, Palmqvist A, Skjold E, Syberg K (2023) Implications of circular textile policies for the future regulation of hazardous substances in textiles in the European Union. Sci Total Environ 896:165153. https://doi.org/10.1016/j.scitotenv.2023.165153
21. Bukhari MA, Carrasco-Gallego R, Ponce-Cueto E (2018) Developing a national programme for textiles and clothing recovery. Waste Manag Res 36:321–331. https://doi.org/10.1177/0734242X18759190
22. Bundgaard AM, Huulgaard RD, Remmen A (2024) Getting the priorities right in material efficiency: from the ecodesign directive to the ecodesign for sustainable products regulation. 2024 Electronics Goes Green 2024+ (EGG). IEEE, Berlin, Germany, pp 1–7
23. Central Pollution Control Board (2025). https://cpcb.nic.in/. Accessed 4 Dec 2025
24. Chapagain A, Hoekstra AY, Savenije HHG, Gautam R (2005) The water footprint of cotton consumption. UNESCO-IHE Institute for Water Education, Delft
25. Chen H, Lin Z, Tan C (2019) Classification of different animal fibers by near infrared spectroscopy and chemometric models. Microchem J 144:489–494. https://doi.org/10.1016/j.microc.2018.10.011
26. Chen X, Memon HA, Wang Y, Marriam I, Tebyetekerwa M (2021) Circular economy and sustainability of the clothing and textile industry. Mater Circ Econ 3:12. https://doi.org/10.1007/s42824-021-00026-2
27. Choudhury K, Tsianou M, Alexandridis P (2024) Recycling of blended fabrics for a circular economy of textiles: separation of cotton, polyester, and elastane fibers. Sustainability 16:6206. https://doi.org/10.3390/su16146206
28. Chowdhury NR, Chowdhury P, Paul SK (2022) Sustainable practices and their antecedents in the apparel industry: a review. Curr Opin Green Sustain Chem 37:100674. https://doi.org/10.1016/j.cogsc.2022.100674
29. Chowdhury NR, Paul SK, Sarker T, Shi Y (2023) Implementing smart waste management system for a sustainable circular economy in the textile industry. Int J Prod Econ 262:108876. https://doi.org/10.1016/j.ijpe.2023.108876
30. CIRCULOSE (2025) CIRCULOSE® is a new material made by recycling cotton from 100% discarded textiles. https://www.circulose.com/en/. Accessed 4 Dec 2025
31. Costa SM, Mazzola PG, Silva JCAR, Pahl R, Pessoa A, Costa SA (2013) Use of sugar cane straw as a source of cellulose for textile fiber production. Ind Crops Prod 42:189–194. https://doi.org/10.1016/j.indcrop.2012.05.028
32. Cristy Dawn (2025). https://christydawn.com/en-in?irclickid=RbmVWPxqxxycRQe3KxRK7WtWUksUkyyZbS3d2Q0&sharedid=&irpid=274622&irgwc=1. Accessed 15 Apr 2025
33. Das S (2011) Life cycle assessment of carbon fiber-reinforced polymer composites. Int J Life Cycle Assess 16:268–282. https://doi.org/10.1007/s11367-011-0264-z

34. Decathlon Sustainability (2025) Is recycled cotton an innovative fabric? https://sustainability.decathlon.com/is-recycled-cotton-an-innovative-fabric. Accessed 4 May 2025
35. Delate K, Heller B, Shade J (2021) Organic cotton production may alleviate the environmental impacts of intensive conventional cotton production. Renew Agric Food Syst 36:405–412. https://doi.org/10.1017/S1742170520000356
36. Din TU, Zaman N, Uddin MN, Qayum N, Qayum S, Arshad M (2025) The circular economy concept in the textile industry. In: Enzymes in textile processing: a climate changes mitigation approach: textile industry, enzymes, and SDGs. Springer Nature, Singapore, pp 331–361
37. Drochytka R, Dvorakova M, Hodna J (2017) Performance evaluation and research of alternative thermal insulation based on waste polyester fibers. Proced Eng 195:236–243. https://doi.org/10.1016/j.proeng.2017.04.549
38. Editorial Team SDG12 (2024) The hidden cost of a cotton T-Shirt. In: Global society news. https://www.globalsociety.earth/post/the-hidden-cost-of-a-cotton-t-shirt. Accessed 14 Apr 2025
39. Eileen Fisher (2025) Thinking in systems Since 1984. https://www.eileenfisher.com/our-brand.html. Accessed 15 Apr 2025
40. El Wazna M, El Fatihi M, El Bouari A, Cherkaoui O (2017) Thermo physical characterization of sustainable insulation materials made from textile waste. J Build Eng 12:196–201. https://doi.org/10.1016/j.jobe.2017.06.008
41. Ellen MacArthur Foundation (2017) A new textiles economy: redesigning fashion's future.
42. EVRNU (2025) A new fiber that can be formed and reformed^{TM}. https://www.evrnu.com/nucycl. Accessed 4 Dec 2025
43. Fahmy A, Mohammed M, Ahmed E, Hassabo A (2024) Recycling and sustainability innovations in the textile industry. J Text Color Pol Sci. https://doi.org/10.21608/jtcps.2024.259259.1300
44. Fei X, Freeman HS, Hinks D (2020) Toward closed loop recycling of polyester fabric: Step 1. Decolorization using sodium formaldehyde sulfoxylate. J Clea Prod 254:120027. https://doi.org/10.1016/j.jclepro.2020.120027
45. Ferreira P, Apolinário A, Forman G (2023) Optimising textile biomaterial selection for sustainable product and circular design: practical guidelines for a greener future. Mater Circ Econ 5:14. https://doi.org/10.1007/s42824-023-00086-6
46. Foulk J, Akin D, Dodd R, Ulven C (2011) Production of flax fibers for biocomposites. In: Kalia S, Kaith BS, Kaur I (eds) Cellulose fibers: bio- and nano-polymer composites. Springer, Berlin Heidelberg, Berlin, Heidelberg, pp 61–95
47. Foulk JA, Chao WY, Akin DE, Dodd RB, Layton PA (2006) Analysis of flax and cotton fiber fabric blends and recycled polyethylene composites. J Polym Environ 14:15–25. https://doi.org/10.1007/s10924-005-8703-1
48. Gholamzad E, Karimi K, Masoomi M (2014) Effective conversion of waste polyester–cotton textile to ethanol and recovery of polyester by alkaline pretreatment. Chem Eng J 253:40–45. https://doi.org/10.1016/j.cej.2014.04.109
49. Global Fashion Agenda (2017) Pulse of the fashion industry-The Boston Consulting Group (BCG). https://globalfashionagenda.org/resource/pulse-of-the-fashion-industry-2017/. Accessed 4 Dec 2025
50. Gonzalez V, Lou X, Chi T (2023) Evaluating environmental impact of natural and synthetic fibers: a life cycle assessment approach. Sustainability 15:7670. https://doi.org/10.3390/su15097670
51. Guo Z, Eriksson M, Motte HDL, Adolfsson E (2021) Circular recycling of polyester textile waste using a sustainable catalyst. J Clean Prod 283:124579. https://doi.org/10.1016/j.jclepro.2020.124579
52. Gupta S, Dangayach GS, Singh AK, Meena ML, Rao PN (2018) Implementation of sustainable manufacturing practices in Indian manufacturing companies. Benchmarking Int J 25:2441–2459. https://doi.org/10.1108/BIJ-12-2016-0186
53. Gupta V, Arora M, Minhas J (2020) Innovating opportunities for fashion brands by using textile waste for better fashion. In: Pandit P, Ahmed S, Singha K, Shrivastava S (eds) Recycling from waste in fashion and textiles, 1st edn. Wiley, pp 101–121

54. Hahn J (2021) Phillip Lim and Charlotte McCurdy adorn couture dress with algae sequins to avoid "reaching for polyester." In: dezeen. https://www.dezeen.com/2021/02/22/charlotte-mccurdy-phillip-lim-algae-sequin-dress-onexone/. Accessed 4 Dec 2025
55. Harane RS, Adivarekar RV (2017) Sustainable processes for pre-treatment of cotton fabric. Text Cloth Sustain 2:2. https://doi.org/10.1186/s40689-016-0012-7
56. Hasan-Al Mamun M, Hossain MA, Zakaria MKA, Hossain MF, Ali R, Yeasmin MS, Rahman L (2017) Effect of different types scouringagainst different types of bleaching process on dyeing of cotton fabric with monochlorotriazine (hot brand) reactive dye. Int J Text Sci 6:128–134
57. Haslinger S, Hummel M, Anghelescu-Hakala A, Määttänen M, Sixta H (2019) Upcycling of cotton polyester blended textile waste to new man-made cellulose fibers. Waste Manage 97:88–96. https://doi.org/10.1016/j.wasman.2019.07.040
58. Haule LV, Carr CM, Rigout M (2016) Preparation and physical properties of regenerated cellulose fibres from cotton waste garments. J Clean Prod 112:4445–4451. https://doi.org/10.1016/j.jclepro.2015.08.086
59. He B, Duan H, Yang W, Yabar H, Song J (2024) Decarbonizing polyamide textile production in China: footprints and mitigation pathways from life cycle perspective. Resour Conserv Recycl 208:107705. https://doi.org/10.1016/j.resconrec.2024.107705
60. Hildebrandt J, Thrän D, Bezama A (2021) The circularity of potential bio-textile production routes: comparing life cycle impacts of bio-based materials used within the manufacturing of selected leather substitutes. J Clean Prod 287:125470. https://doi.org/10.1016/j.jclepro.2020.125470
61. Holaday C (2024) First-in-the-nation law forces certain stores to take old clothes back for free as officials take aim at 'fast fashion.' In: The U.S. Sun. https://www.the-sun.com/money/12590037/california-bill-fast-fashion/. Accessed 4 Nov 2025
62. Holland C, Perzon A, Cassland PRC, Jensen JP, Langebeck B, Sørensen OB, Whale E, Hepworth D, Plaice-Inglis R, Moestrup Ø, Ulvskov P, Jørgensen B (2019) Nanofibers produced from agro-industrial plant waste using entirely enzymatic pretreatments. Biomacromol 20:443–453. https://doi.org/10.1021/acs.biomac.8b01435
63. Hörner Bussolo G, Bressanelli G, Visintin F, Saccani N (2024) Toward the European Union 2030 strategy for textiles: a review. In: Thürer M, Riedel R, Von Cieminski G, Romero D (eds) Advances in production management systems. Production management systems for volatile, uncertain, complex, and ambiguous environments. Springer Nature Switzerland, Cham, pp 382–397
64. Hu Y, Du C, Leu S-Y, Jing H, Li X, Lin CSK (2018) Valorisation of textile waste by fungal solid state fermentation: an example of circular waste-based biorefinery. Resour Conserv Recycl 129:27–35. https://doi.org/10.1016/j.resconrec.2017.09.024
65. Huijbregts MAJ, Steinmann ZJN, Elshout PMF, Stam G, Verones F, Vieira M, Zijp M, Hollander A, Van Zelm R (2017) ReCiPe2016: a harmonised life cycle impact assessment method at midpoint and endpoint level. Int J Life Cycle Assess 22:138–147. https://doi.org/10.1007/s11367-016-1246-y
66. IEA Bioenergy (2025) New publication–advanced biofuels–potential for cost reduction. https://www.ieabioenergy.com/blog/publications/new-publication-advanced-biofuels-potential-for-cost-reduction/. Accessed 4 Dec 2025
67. Islam MM, Perry P, Gill S (2021) Mapping environmentally sustainable practices in textiles, apparel and fashion industries: a systematic literature review. JFMM 25:331–353. https://doi.org/10.1108/JFMM-07-2020-0130
68. Jeff Sloan (2019) LeMond Carbon audits rapid-oxidation carbon fiber technology. In: CW compositesWorld. https://www.compositesworld.com/news/lemond-carbon-audits-rapid-oxidation-carbon-fiber-technology. Accessed 4 Dec 2025
69. Jeihanipour A, Taherzadeh MJ (2009) Ethanol production from cotton-based waste textiles. Biores Technol 100:1007–1010. https://doi.org/10.1016/j.biortech.2008.07.020
70. Jhanji Dhir Y (2022) Natural fibers: the sustainable alternatives for textile and non-textile applications. In: Jeon H-Y (ed) Natural fiber. IntechOpen

71. Jia F, Yin S, Chen L, Chen X (2020) The circular economy in the textile and apparel industry: a systematic literature review. J Clean Prod 259:120728. https://doi.org/10.1016/j.jclepro.2020.120728
72. Jiang X, Bai Y, Chen X, Liu W (2020) A review on raw materials, commercial production and properties of lyocell fiber. J Bioresour Bioprod 5:16–25. https://doi.org/10.1016/j.jobab.2020.03.002
73. Jiu Z, Sethi MH (2025) Is upcycling the best approach for designers to contribute to sustainable textile industry? TEXTILE 23:276–288. https://doi.org/10.1080/14759756.2025.2465167
74. Jose S, Salim R, Ammayappan L (2016) An overview on production, properties, and value addition of pineapple leaf fibers (PALF). J Nat Fibers 13:362–373. https://doi.org/10.1080/15440478.2015.1029194
75. Jozala AF, De Lencastre-Novaes LC, Lopes AM, De Carvalho S-E, Mazzola PG, Pessoa-Jr A, Grotto D, Gerenutti M, Chaud MV (2016) Bacterial nanocellulose production and application: a 10-year overview. Appl Microbiol Biotechnol 100:2063–2072. https://doi.org/10.1007/s00253-015-7243-4
76. Juanga-Labayen J, Yanac K, Yuan Q (2021) Effect of substrate-to-inoculum ratio on anaerobic digestion of treated and untreated cotton textile waste. Int J Environ Sci Technol 18:287–296. https://doi.org/10.1007/s13762-020-02831-9
77. Kalayci E, Gokmen Isanc E, Avinc O (2023) Tillandsia usneoides L. (Spanish Moss) air plant and its important potential for sustainable technical textile applications. In: Muthu SS (ed) Novel sustainable raw material alternatives for the textiles and fashion industry. Springer Nature Switzerland, Cham, pp 35–55
78. Kelly C (2025) Closing the loop: an exploration of the challenges and opportunities in circular economy adoption in the outdoor apparel industry
79. Khabbaz BG (2010) Life cycle energy use and greenhouse gas emissions of Australian cotton: impact of farming systems. Doctoral Thesis, University of Southern Queensland.
80. Khan A, Iftikhar K, Mohsin M, Ubaidullah M, Ali M, Mueen A (2022) Banana agro-waste as an alternative to cotton fibre in textile applications. Yarn to fabric: an ecofriendly approach. Ind Crop Prod 189:115687. https://doi.org/10.1016/j.indcrop.2022.115687
81. Kozlowski R, Muzyczek M (2023) Hemp, flax and other plant fibres. In: Sustainable fibres for fashion and textile manufacturing. Elsevier, pp 75–93
82. Krywalski-Santiago J (2024) Navigating sustainable transformation in the fashion industry: the role of circular economy and ethical consumer behavior. J Intercult Manag 16:5–36. https://doi.org/10.2478/joim-2024-0009
83. Kumar S, Singh K (2020) Tribological behaviour of fibre-reinforced thermoset polymer composites: a review. Proc Inst Mech Eng Part L J Mater Design Appl 234:1439–1449. https://doi.org/10.1177/1464420720941554
84. Kumar V, Kushwaha PK, Kumar R (2011) Impedance-spectroscopy analysis of oriented and mercerized bamboo fiber-reinforced epoxy composite. J Mater Sci 46:3445–3451. https://doi.org/10.1007/s10853-011-5249-6
85. Kumartasli S, Avinc O (2020) Important step in sustainability: polyethylene terephthalate recycling and the recent developments. In: Muthu SS, Gardetti MA (eds) Sustainability in the textile and apparel industries. Springer International Publishing, Cham, pp 1–19
86. Kunchimon SZ, Tausif M, Goswami P, Cheung V (2019) Polyamide 6 and thermoplastic polyurethane recycled hybrid fibres via twin-screw melt extrusion. J Polym Res 26:162. https://doi.org/10.1007/s10965-019-1827-0
87. Kurien RA, Maria Anil M, Sharan Mohan SL, Anna Thomas J (2023) Natural fiber composites as sustainable resources for emerging applications-a review. Mater Today Proceed S2214785323022642. https://doi.org/10.1016/j.matpr.2023.04.363
88. Kwakwa PA (2025) Decarbonizing the textiles and clothing industry in South Africa: the role of renewable energy. RJTA. https://doi.org/10.1108/RJTA-09-2024-0162
89. La Rosa AD, Cozzo G, Latteri A, Recca A, Björklund A, Parrinello E, Cicala G (2013) Life cycle assessment of a novel hybrid glass-hemp/thermoset composite. J Clean Prod 44:69–76. https://doi.org/10.1016/j.jclepro.2012.11.038

90. La Rosa AD, Grammatikos SA (2019) Comparative life cycle assessment of cotton and other natural fibers for textile applications. Fibers 7:101. https://doi.org/10.3390/fib7120101
91. Le K (2018) Textile recycling technologies, colouring and finishing methods. Solid Waste Services, Metro Vancouver, Vancouver
92. Leal Filho W, Perry P, Heim H, Dinis MAP, Moda H, Ebhuoma E, Paço A (2022) An overview of the contribution of the textiles sector to climate change. Front Environ Sci 10:973102. https://doi.org/10.3389/fenvs.2022.973102
93. Li X, Zhang M, Luo J, Zhang S, Yang X, Igalavithana AD, Ok YS, Tsang DCW, Lin CSK (2019) Efficient succinic acid production using a biochar-treated textile waste hydrolysate in an in situ fibrous bed bioreactor. Biochem Eng J 149:107249. https://doi.org/10.1016/j.bej.2019.107249
94. Liu D, Song J, Anderson DP, Chang PR, Hua Y (2012) Bamboo fiber and its reinforced composites: structure and properties. Cellulose 19:1449–1480. https://doi.org/10.1007/s10570-012-9741-1
95. Loo S-L, Yu E, Hu X (2023) Tackling critical challenges in textile circularity: a review on strategies for recycling cellulose and polyester from blended fabrics. J Environ Chem Eng 11:110482. https://doi.org/10.1016/j.jece.2023.110482
96. Lu JJ, Hamouda H (2014) Current status of fiber waste recycling and its future. AMR 878:122–131. https://doi.org/10.4028/www.scientific.net/AMR.878.122
97. Ma Y, Hummel M, Määttänen M, Särkilahti A, Harlin A, Sixta H (2016) Upcycling of waste paper and cardboard to textiles. Green Chem 18:858–866. https://doi.org/10.1039/C5GC01679G
98. Maino A, Janszen G, Di Landro L (2019) Glass/epoxy and hemp/bio based epoxy composites: manufacturing and structural performances. Polym Compos 40.https://doi.org/10.1002/pc.24973
99. Maiti R (2025) The environmental impact of fast fashion, explained. In: EARTH.ORG. https://earth.org/fast-fashions-detrimental-effect-on-the-environment/. Accessed 14 Apr 2025
100. Malai (2025). https://malai.eco/. Accessed 15 Apr 2025
101. Manaia JP, Manaia AT, Rodriges L (2019) Industrial hemp fibers: an overview. Fibers 7:106. https://doi.org/10.3390/fib7120106
102. Martin AR, Martins MA, Da Silva ORRF, Mattoso LHC (2010) Studies on the thermal properties of sisal fiber and its constituents. Thermochim Acta 506:14–19. https://doi.org/10.1016/j.tca.2010.04.008
103. MATE (2025) Dress Clean®. http://matethelabel.com/pages/about. Accessed 4 Dec 2025
104. Moazzem S, Crossin E, Daver F, Wang L (2021) Assessing environmental impact reduction opportunities through life cycle assessment of apparel products. Sustain Prod Consum 28:663–674. https://doi.org/10.1016/j.spc.2021.06.015
105. Mochane MJ, Mokhena TC, Mokhothu TH, Mtibe A, Sadiku ER, Ray SS, Ibrahim ID, Daramola OO (2019) Recent progress on natural fiber hybrid composites for advanced applications: a review. Express Polym Lett 13:159–198. https://doi.org/10.3144/expresspolymlett.2019.15
106. More AP (2022) Flax fiber–based polymer composites: a review. Adv Compos Hybrid Mater 5:1–20. https://doi.org/10.1007/s42114-021-00246-9
107. Mushtaq A (2025) Advancements in sustainable fabric production: a review of technologies and future prospects. Bull Eng Sci Technol Ind 3:16–18
108. Muthu SS (2015) Sustainable UV-protective apparel textile. In: Handbook of sustainable apparel production, 0 ed. CRC Press, pp 128–155
109. Muthukumarana TT, Karunathilake HP, Punchihewa HKG, Manthilake MMID, Hewage KN (2018) Life cycle environmental impacts of the apparel industry in Sri Lanka: analysis of the energy sources. J Clean Prod 172:1346–1357. https://doi.org/10.1016/j.jclepro.2017.10.261
110. MYCOWORKS (2022) Grow the future of materials. https://www.mycoworks.com/fine-mycelium-an-advanced-materials-platform. Accessed 14 Apr 2025
111. Nanollose (2025). https://nanollose.com/. Accessed 4 Dec 2025

112. Natural Fiber Market (2025–2030) (2025) Size, share & trends analysis report by plant-based fiber (Cotton, Linen, Jute, Hemp), By animal-based fiber (Wool, Silk), By region (North America, Europe, Asia Pacific, C&SA, MEA), And segment forecasts
113. Nayak R (2023) Sustainable fibres for fashion and textile manufacturing. Woodhead publishing, Cambridge (Ma)
114. Nayak R, Cleveland D, Tran G, Joseph F (2024) Potential of bacterial cellulose for sustainable fashion and textile applications: a review. J Mater Sci 59:6685–6710. https://doi.org/10.1007/s10853-024-09577-6
115. Neira J (2021) This dress has been crafted using algae-based sequins & carbon-neutral fabric. In: Designboom. https://www.designboom.com/design/phillip-lim-charlotte-mccurdy-algae-sequin-dress-02-28-2021/. Accessed 4 Nov 2025
116. Niinimäki K, Peters G, Dahlbo H, Perry P, Rissanen T, Gwilt A (2020) The environmental price of fast fashion. Nat Rev Earth Environ 1:189–200. https://doi.org/10.1038/s43017-020-0039-9
117. Östlund A, Syren P-O, Jonsson C, Ribitsch D, Syren M (2017) Re: mix-separation and recycling of textile waste fiber blends. Mistra Future Fashion, Borås, Sweden
118. Pakkanen V (2024) Polyester production: transitioning from fossil fuels to sustainable alternatives. Master's thesis, Lappeenranta–Lahti University of Technology
119. Palme A, Peterson A, De La Motte H, Theliander H, Brelid H (2017) Development of an efficient route for combined recycling of PET and cotton from mixed fabrics. Text Cloth Sustain 3:4. https://doi.org/10.1186/s40689-017-0026-9
120. Panthapulakkal S, Sain M (2007) Studies on the water absorption properties of short hemp—glass fiber hybrid polypropylene composites. J Compos Mater 41:1871–1883. https://doi.org/10.1177/0021998307069900
121. Peijs T (2000) Natural fiber based composites. Mater Technol 15:281–285. https://doi.org/10.1080/10667857.2000.11752892
122. Pender K, Romoli F, Martin Rodes FA, Fuller J, Zeolla M (2025) Future strategies for decarbonisation of carbon fibre products: a roadmap to net zero 2050. J Clean Prod 486:144525. https://doi.org/10.1016/j.jclepro.2024.144525
123. Peters G, Svanström M, Roos S, Sandin G, Zamani B (2015) Carbon footprints in the textile industry. In: Handbook of life cycle assessment (LCA) of textiles and clothing. Elsevier, pp 3–30
124. Pickering KL, Efendy MGA, Le TM (2016) A review of recent developments in natural fibre composites and their mechanical performance. Compos A Appl Sci Manuf 83:98–112. https://doi.org/10.1016/j.compositesa.2015.08.038
125. Popescu C, Dissanayake H, Mansi E, Stancu A (2024) Eco breakthroughs: sustainable materials transforming the future of our planet. Sustainability 16:10790. https://doi.org/10.3390/su162310790
126. PR Newswire (2019) Södra's breakthrough will enable large-scale textile recycling. https://www.prnewswire.com/in/news-releases/sodra-s-breakthrough-will-enable-large-scale-textile-recycling-850423884.html. Accessed 4 May 2025
127. Prenzel TM, Hohmann A, Prescher T, Angerer K, Wehner D, Ilg R, Von Reden T, Drechsler K, Albrecht S (2023) Bringing light into the dark—overview of environmental impacts of carbon fiber production and potential levers for reduction. Polymers 16:12. https://doi.org/10.3390/polym16010012
128. Radzi AM, Zaki S, Hassan M, Ilyas RA, Jamaludin K, Daud M, Aziz S (2022) Bamboo-fiber-reinforced thermoset and thermoplastic polymer composites: a review of properties, fabrication, and potential applications. Polymers 14:1387. https://doi.org/10.3390/polym14071387
129. Rajeshkumar G, Devnani GL, Maran JP, Sanjay MR, Siengchin S, Al-Dhabi NA, Ponmurugan K (2021) Characterization of novel natural cellulosic fibers from purple bauhinia for potential reinforcement in polymer composites. Cellulose 28:5373–5385. https://doi.org/10.1007/s10570-021-03919-2

130. Ramírez-Escamilla HG, Martínez-Rodríguez MC, Padilla-Rivera A, Domínguez-Solís D, Campos-Villegas LE (2024) Advancing toward sustainability: a systematic review of circular economy strategies in the textile industry. Recycling 9:95. https://doi.org/10.3390/recycling9050095
131. Ranalli P, Venturi G (2004) Hemp as a raw material for industrial applications. Euphytica 140:1–6. https://doi.org/10.1007/s10681-004-4749-8
132. Rao F, Ji Y, Li N, Zhang Y, Chen Y, Yu W (2020) Outdoor bamboo-fiber-reinforced composite: influence of resin content on water resistance and mechanical properties. Constr Build Mater 261:120022. https://doi.org/10.1016/j.conbuildmat.2020.120022
133. Rastogi M, Shrivastava S (2017) Recent advances in second generation bioethanol production: an insight to pretreatment, saccharification and fermentation processes. Renew Sustain Energy Rev 80:330–340. https://doi.org/10.1016/j.rser.2017.05.225
134. Rathinamoorthy R, Kiruba T (2022) Bacterial cellulose-a potential material for sustainable eco-friendly fashion products. J Nat Fibers 19:3275–3287. https://doi.org/10.1080/15440478.2020.1842841
135. Malabadi RB, Kolkar KP, Chalannavar RK (2023) Industrial Cannabis sativa: role of hemp (fiber type) in textile industries. World J Bio Pharm Health Sci 16:001–014. https://doi.org/10.30574/wjbphs.2023.16.2.0450
136. Reilly M, Dinsdale R, Guwy A (2015) Enhanced biomethane potential from wheat straw by low temperature alkaline calcium hydroxide pre-treatment. Biores Technol 189:258–265. https://doi.org/10.1016/j.biortech.2015.03.150
137. Reiniati I, Hrymak AN, Margaritis A (2017) Recent developments in the production and applications of bacterial cellulose fibers and nanocrystals. Crit Rev Biotechnol 37:510–524. https://doi.org/10.1080/07388551.2016.1189871
138. Remington C (2019) Textile exchange debut for Spinnova's straw fibre. In: Ecotextile News. https://www.ecotextile.com/2019101637687/news/materials-production-news/textile-exchange-debut-for-spinnova-s-straw-fibre/. Accessed 4 Dec 2025
139. Ribul M, Lanot A, Tommencioni Pisapia C, Purnell P, McQueen-Mason SJ, Baurley S (2021) Mechanical, chemical, biological: moving towards closed-loop bio-based recycling in a circular economy of sustainable textiles. J Clean Prod 326:129325. https://doi.org/10.1016/j.jclepro.2021.129325
140. Roos S, Sandin G, Peters G, Spak B, Bour LS, Perzon E, Jönsson C (2019) White paper on textile recycling. https://doi.org/10.13140/RG.2.2.31018.77766
141. Salmela J, Widmaier T, Kuosmanen P, Kiviluoma P, Liukkonen J, Koskinen H, Lehto J Method and apparatus for producing fibre yarn
142. Sarda P, Hanan JC, Lawrence JG, Allahkarami M (2022) Sustainability performance of polyethylene terephthalate, clarifying challenges and opportunities. J Polym Sci 60:7–31. https://doi.org/10.1002/pol.20210495
143. Sathishkumar GK, Ibrahim M, Mohamed Akheel M, Rajkumar G, Gopinath B, Karpagam R, Karthik P, Martin Charles M, Gautham G, Gowri Shankar G (2022) Synthesis and mechanical properties of natural fiber reinforced epoxy/polyester/polypropylene composites: a review. J Nat Fibers 19:3718–3741. https://doi.org/10.1080/15440478.2020.1848723
144. Schmidt A, Watson D, Roos S, Askham C, Poulsen PB (2016) Gaining benefits from discarded textiles. Nordic Council of Ministers
145. Seidu RK, Eghan B, Acquaye R (2024) A review of circular fashion and bio-based materials in the fashion industry. CircEconSust 4:693–715. https://doi.org/10.1007/s43615-023-00303-z
146. Şener Fidan F, Kızılkaya Aydoğan E, Uzal N (2021) A comparative life cycle assessment of conventional and organic cotton in denim fabric. Thessaloniki, Greece, pp 1–2
147. Şener Fidan F, Kızılkaya Aydoğan E, Uzal N (2022) The impact of organic cotton use and consumer habits in the sustainability of jean production using the LCA approach. Environ Sci Pollut Res 30:8853–8867. https://doi.org/10.1007/s11356-022-22872-6
148. Shahzad A (2012) Hemp fiber and its composites–a review. J Compos Mater 46:973–986. https://doi.org/10.1177/0021998311413623

149. Sharma S, Jahanzaib M, Lee J, Park D (2025) Global legislation, schemes and standards to control environmental concern for textile-generated waste. In: Sharma P, Sahay SS (eds) Nanotechnology-assisted recycling of textile waste, 1st edn. Wiley, pp 435–459
150. Sherwood J (2020) Closed-loop recycling of polymers using solvents: remaking plastics for a circular economy. Johns Matthey Technol Rev 64:4–15. https://doi.org/10.1595/205651319X15574756736831
151. Shirvanimoghaddam K, Motamed B, Ramakrishna S, Naebe M (2020) Death by waste: fashion and textile circular economy case. Sci Total Environ 718:137317. https://doi.org/10.1016/j.scitotenv.2020.137317
152. Shuhua W, Xiaoying Y, Xiaogang C, Wensheng H, Mei N (2020) Recycling of cotton fibers separated from the waste blend fabric. J Nat Fibers 17:520–531. https://doi.org/10.1080/15440478.2018.1503130
153. Singha K, Pandit P, Maity S, Sharma SR (2021) Harmful environmental effects for textile chemical dyeing practice. In: Green chemistry for sustainable textiles. Elsevier, pp 153–164
154. Subramanian K, Chopra SS, Cakin E, Li X, Lin CSK (2020) Environmental life cycle assessment of textile bio-recycling–valorizing cotton-polyester textile waste to pet fiber and glucose syrup. Resour Conserv Recycl 161:104989. https://doi.org/10.1016/j.resconrec.2020.104989
155. Sustainable Manufacturing Expo (2025) Innovations in eco-friendly textile production: expert perspectives. https://www.sustainablemanufacturingexpo.com/en/articles/eco-friendly-textile-innovations.html. Accessed 4 Oct 2025
156. TED (2011) TED 2011: fermented frocks, the new couture. In: WIRED. https://www.wired.com/2011/03/fermented-frocks/. Accessed 4 Nov 2025
157. Teli M, Valia S, Kadu K, Dahale M (2015) Upcycling of textile materials. Bangkok, Thailand, pp 164–167
158. Tencel (2025) Our fibers: Tencel weaves together the story of nature and people. In: Tencel. https://www.tencel.com/fibers. Accessed 4 Dec 2025
159. Thakur VK, Thakur MK (2014) Processing and characterization of natural cellulose fibers/thermoset polymer composites. Carbohyd Polym 109:102–117. https://doi.org/10.1016/j.carbpol.2014.03.039
160. The Good Trade (2025) 15 Most eco-conscious clothing brands in 2025. https://www.thegoodtrade.com/features/eco-friendly-clothing-brands/. Accessed 15 Apr 2025
161. Tian W, Huang K, Zhu C, Sun Z, Shao L, Hu M, Feng X (2022) Recent progress in biobased synthetic textile fibers. Front Mater 9:1098590. https://doi.org/10.3389/fmats.2022.1098590
162. Todor MP, Bulei C, Kiss I, Alexa V (2019) Recycling of textile wastes into textile composites based on natural fibres: the valorisation potential. IOP Conf Ser Mater Sci Eng 477:012004. https://doi.org/10.1088/1757-899X/477/1/012004
163. UN Environment Programme (2025) Unsustainable fashion and textiles in focus for International Day of Zero Waste 2025. https://www.unep.org/news-and-stories/press-release/unsustainable-fashion-and-textiles-focus-international-day-zero#:~:text=Every%20year%2C%2092%20million%20tonnes,2023%20made%20from%20recycled%20sources. Accessed 14 Apr 2025
164. Unal F, Avinc O, Yavas A (2022) Production of sustainable banana fibers from agricultural wastes and their properties. In: Muthu SS (ed) Sustainable approaches in textiles and fashion. Springer Nature Singapore, Singapore, pp 157–193
165. UNEP (2025) The UNEP textile initiative. https://www.unep.org/topics/chemicals-and-pollution-action/circularity-sectors/sustainable-and-circular-textiles. Accessed 4 Dec 2025
166. Vecchiato S, Skopek L, Jankova S, Pellis A, Ipsmiller W, Aldrian A, Mueller B, Herrero Acero E, Guebitz GM (2018) Enzymatic recycling of high-value phosphor flame-retardant pigment and glucose from rayon fibers. ACS Sustain Chem Eng 6:2386–2394. https://doi.org/10.1021/acssuschemeng.7b03840
167. Wan PKF, Jiang S (2025) Enabling a dynamic information flow in digital product passports during product use phase: a literature review and proposed framework. Sustain Prod Consum 54:362–374. https://doi.org/10.1016/j.spc.2025.01.014

168. Wang H, Siddiqui MQ, Memon H (2020) Physical structure, properties and quality of cotton. In: Cotton science and processing technology. Springer Singapore, Singapore
169. Wang Y, Zhu W, Wan B, Meng Z, Han B (2021) Hygrothermal ageing behavior and mechanism of carbon nanofibers modified flax fiber-reinforced epoxy laminates. Compos A Appl Sci Manuf 140:106142. https://doi.org/10.1016/j.compositesa.2020.106142
170. Warwas I, Podgórniak-Krzykacz A, Przywojska J, Kozar Ł (2021) Going green and socially responsible–textile industry in transition to sustainability and a circular economy. Fibres & Textiles East Eur 29:8–18. https://doi.org/10.5604/01.3001.0014.7782
171. Wossink A, Denaux ZS (2006) Environmental and cost efficiency of pesticide use in transgenic and conventional cotton production. Agric Syst 90:312–328. https://doi.org/10.1016/j.agsy.2006.01.004
172. Xia L, Chun D (2012) The antibacterial performance of natural bamboo fiber and its influencing factors. In: Materials science, environmental science. Beijing, China, pp 1–8
173. Yeh C-H, Yang T-C (2020) Utilization of waste bamboo fibers in thermoplastic composites: influence of the chemical composition and thermal decomposition behavior. Polymers 12:636. https://doi.org/10.3390/polym12030636
174. Yousef S, Tatariants M, Tichonovas M, Kliucininkas L, Lukošiūtė S-I, Yan L (2020) Sustainable green technology for recovery of cotton fibers and polyester from textile waste. J Clean Prod 254:120078. https://doi.org/10.1016/j.jclepro.2020.120078
175. Yousif BF, Orupabo C, Azwa ZN (2012) Characteristics of Kenaf fiber immersed in different solutions. J Nat Fibers 9:207–218. https://doi.org/10.1080/15440478.2012.733149
176. Zavagno A (2021) Upcycling as a design strategy to innovate fashion processes and create a local value chain with the support ofdigital technologies: the Offbeat District case. Master's thesis, Politecnico di Milano
177. Zhang L, Leung MY, Boriskina S, Tao X (2022) Advancing life cycle sustainability of textiles through technological innovations. Nat Sustain 6:243–253. https://doi.org/10.1038/s41893-022-01004-5
178. Zhong C (2020) Industrial-scale production and applications of bacterial cellulose. Front Bioeng Biotechnol 8:605374. https://doi.org/10.3389/fbioe.2020.605374

Advancing Sustainable Practices in Textiles: Exploring Innovative Resource Utilization

Suraj Prakash, Radha, and Sunil Puri

Abstract The global textile industry is still working under a linear system known as take, make and dispose. This system is associated with environmental damage and different social problems. The current pattern leads to high use of natural resources, pollution in land and water bodies, and a huge amount of textile waste. Because of this, there is a strong need to move towards circular and sustainable methods to reduce these harmful effects. This book chapter discusses how sustainable practices have gradually evolved in the textile industry and focus is given on innovative resource utilization as a key approach towards achieving sustainability. Under material innovation, several practices such as utilization of agricultural residues, development of textiles from fungal mycelium and marine algae, chemical recycling of post-consumer textile waste, and production of biodegradable polymers are discussed. Process-related innovations are highlighted with examples like resource-saving techniques in manufacturing including supercritical CO_2 dyeing and digital textile printing, along with detailed study of recycling technologies. In addition, this book chapter also discusses the application of Life Cycle Assessment (LCA) as framework for evaluating sustainability and critically highlights major ongoing debates in the industry such as whether bio-based or recycled synthetic fibers are more sustainable. However, some knowledge gaps were also found in literature including lack of studies on integrated systems approach, limited understanding about social impacts of new technologies, and insufficient data available from developing countries perspective. For future direction, more emphasis should be given on interdisciplinary research to develop circular textile economy that is not only technologically feasible but also practically scalable and socially just.

S. Prakash · Radha (✉) · S. Puri
School of Biological and Environmental Sciences, Shoolini University, Solan, Himachal Pradesh, India
e-mail: radhuchauhan7002@gmail.com

S. Prakash
e-mail: surajpandiar75@gmail.com

S. Puri
e-mail: sunilpuri@shooliniuniversity.com

S. S. Muthu (ed.), *Resource Mixes in Textiles*, Sustainable Textiles: Production, Processing, Manufacturing & Chemistry, https://doi.org/10.1007/978-3-032-10921-7_6

Keywords Stainability · Resource utilization · Plant extracts · Circular economy · Eco-friendly textiles

1 Introduction

The textile and apparel sector is considered one of the largest industries globally, offering employment to around 20 millions of people and producing a large number of garments every year [1]. Over the years, a business approach called fast fashion has become dominant, but it has also led to serious environmental and social problems. This system mainly works on a linear method of production and consumption often referred to as "take, make, dispose," where clothes are designed to meet short-term trends and replaced quickly to maintain consumer demand [2]. As a result, the negative effects of this system are now widely studied and known. Reports suggest that this industry contributes to nearly 6–8% of the world's carbon emissions, releasing around 1.7 billion tons of CO_2 every year [3]. It accounts for approximately 20% of global industrial water pollution, primarily from dyeing and finishing processes that release toxic effluents into ecosystems, and second-largest consumer of the world's freshwater resources [4, 5]. In 2015 alone, the sector consumed 79 billion cubic meters of water and these figures are projected to surge by over 50% by 2030 if the current trajectory continues [6]. The end of life stage of this linear system is equally calamitous. Globally, an estimated 75% of all textile waste is disposed of in landfills or incinerated, while a mere 1% is recycled back into new clothing, representing a massive loss of material and economic value [7]. This waste crisis is compounded by the pervasive issue of microplastic pollution, where washing synthetic garments releases vast quantities of non-biodegradable fibers into waterways and oceans [8]. Beyond the ecological fallout, the fast fashion model's relentless pressure on cost and speed has created a significant social deficit. Production is frequently outsourced to countries with lower labor costs, leading to the exploitation of workers in long, fragmented, and often opaque supply chains characterized by unsafe working conditions and inadequate compensation [9]. These statistics are not merely data points; they are indicators of a systemic failure, a model that externalizes its true costs onto the planet and its people, creating an urgent and undeniable imperative for a fundamental resource revolution.

The response from academic and industrial sectors towards the ongoing textile crisis has changed gradually in the last few decades. The idea of sustainability became more prominent in the late 1980s, mainly after the Brundtland Report introduced the concept of development that meets present needs without affecting the needs of future generations [10]. Later, John Elkington (1998) proposed the "Triple Bottom Line" concept, which explained sustainability in terms of three main areas: social equity (People), environmental care (Planet), and economic growth (Profit) [11]. This approach showed better use of available resources, such as lowering water use in dyeing processes or improving working conditions in textile industries. In the early 2000s, research in the field of sustainable fashion started to increase, especially after 2010. New academic journals such as *Fashion and Textiles* (2014) and *Sustainable Production and Consumption* (2015) were started to publish studies focused on this topic [12]. Reports and findings showed that making small changes in current system is not enough to reduce the rising impact of textile sector on the environment [6]. As a result, the focus started shifting from improving the current practices to adopting a new concept known as Circular Economy (CE). The concept of CE is based on three main principles: avoiding waste and pollution from the beginning, keeping products and materials in use for a longer time, and helping nature to recover [13]. This model does not only aim at reducing environmental damage but focuses on creating systems that can restore and support nature in future. In recent years, studies have shown this change where earlier research mostly focused on consumer choices related to sustainable clothing. Currently, research expanded toward circular models in business, changes in supply chains, and responsibility shown by industries towards society and the environment [14]. This book chapter examine the development of sustainability in the textile industry by first providing the historical, economic, and ecological background, contrasting linear take, make, dispose model with earlier efficiency reforms that were not sufficient. Further, literature on fast fashion, rising carbon and water footprints, global textile waste, and labor issues is reviewed to show how attention has shifted toward sustainability and circularity. Methodological approaches commonly used in this field are then showed such as life cycle assessments, social assessments, fiber testing, and process evaluations including chemical recycling and waterless dyeing, with a focus on both their contributions and limitations. Findings from earlier studies are synthesized in two major areas: material innovations, such as recovery of natural fibers, bio fabricated textiles, and biodegradable polymers; and process innovations, such as textile recycling, energy optimization, and digital dyeing. Key challenges are identified at the technical level, including fiber quality and recycling efficiency, as well as at the economic, logistical, and regulatory levels, where high costs, weak collection systems, and incomplete assessments remain obstacles. Final section outlines future directions and opportunities, presenting the chapter as a link between present knowledge and the innovations needed to build a circular and resilient textile economy.

2 Next Generation of Material Feedstocks

The base of a sustainable textile system depends mainly on the type of raw materials used. Earlier, the textile industry was mostly dependent on virgin fibers such as traditional cotton and synthetic fibers made from fossil fuels, which has become a major reason for environmental damage [6]. To deal with this issue, researchers have started working on different types of new raw materials. These include using waste materials in a better way and also making new fibers with the help of biotechnology.

2.1 Valorizing Lignocellulosic Waste

Among different approaches, the use of agricultural waste for making textile fibers is getting more attention. It is a way to turn waste into a useful resource. Every year, a large amount of agricultural biomass like leaves, husks, and stalks is produced as a leftover from food crops. Around 140 gigatons of such biomass is generated yearly, and most of it remains unused [15]. Scientists are now studying how to take out useful fibers from these materials. Fibers can be obtained from plant waste like banana stem, pineapple leaves, flax, jute, and hemp for textile use [16]. Banana fiber (also known as musa fiber), for example, has gained significant attention for its exceptional properties. It is considered one of the world's strongest natural fibers, exhibiting a high tensile strength of approximately 458 MPa, comparable to or exceeding that of many conventional natural fibers. Furthermore, banana fibers are biodegradable, recyclable, and possess potential to enhance fire resistance properties of cotton fabric [17, 18]. Studies showed different methods to separate cellulosic fibers from the surrounding matrix of lignin and hemicellulose. Some of the commonly used methods include chemical treatment, mechanical processes, biological approaches, and combination of these. In chemical methods, substances like sodium hydroxide or acids are used to remove non-cellulosic parts. Mechanical methods involve physical processes such as grinding, refining, or steam-based treatments to break the structure of biomass [19]. While in biological methods, specific enzymes such as pectinase and xylanase are used for selective separation in a mild way [20]. In some cases, a combination of chemical and mechanical techniques is applied to improve fiber quality and yield [21]. However, even though agricultural residues have high potential, there are still many challenges in their use at industrial level. Primary hurdle is achieving consistency in fiber quality and mechanical properties, which can vary significantly depending on the source plant, growing conditions, and extraction method used [22]. Furthermore, scaling up extraction technologies from the laboratory to industrial production while maintaining economic viability remains a significant barrier. The logistics of collecting, transporting, and processing what is often a diffuse and seasonally variable waste stream also present considerable operational complexities [23].

2.2 Biofabricated Textiles

Moving ahead from the traditional method of harvesting fibers from plants, a new direction is seen where materials are grown directly from living organisms. This method uses biotechnology to develop textiles that have unique properties and support sustainable use. Among the living organisms used in this process, fungi (especially mycelium) and algae are the most studied. Mycelium is the root-like structure of fungi made of fine threads. It grows on different organic materials such as crop waste or even textile waste. During its growth, it binds the material together and forms a compact, interlinked structure in three dimensions [24]. Through controlled cultivation, this mycelial mass can be processed into materials that mimic the properties of leather soft, flexible, and durable, yet fully biodegradable at the end of their life (Lee et al. 2020). Several companies have already begun to pioneer commercial products, signaling a transition from speculative research to market reality [25]. Algae based textile is an emerging area in the field of biofabrication which offer various sustainability benefits. Algae are known for their fast growth and can be easily cultivated in aquatic habitats including seawater. They need less land and freshwater and also absorb carbon dioxide during growth through photosynthesis [26]. The main polymer obtained from brown algae is alginate, which is widely used for making fiber. Alginate can be converted into fiber using wet spinning method, but in raw form it shows low strength and poor resistance to water which makes it less suitable for textile purpose [27]. To overcome this issue, recent studies have used natural cross-linking agents to improve its properties. It was found that cross-linking alginate with compounds such as chitosan, ferulic acid or citric acid can help in improving the mechanical strength of fiber. For instance, fiber made by using ferulic acid shows good stiffness and strength similar to wool, while fiber cross-linked with citric acid shows more elasticity and properties like nylon [27]. Beyond fibers, algae are also a rich source of natural pigments, which can be used to create bio-based dyes and coatings, enabling the production of fully bio-based and colored textile products [28]. The primary barriers to widespread adoption are the currently high production costs relative to conventional fibers and the technical challenges associated with achieving consistent raw material quality and optimizing fiber performance for large-scale manufacturing.

2.3 Advanced Regeneration

While development of new biomaterials is receiving growing attention, one of the key aspects of circular economy is the proper management of textile waste that has already been produced. In recent years, technologies are being developed to convert post-consumer textile waste into usable fibers by breaking down the material chemically and converting it into fibers that are similar in quality to the original ones [29]. This process is important to establish a system where commonly used textile fibers can be reused without generating new waste. Earlier studies were mostly focused on clean and unused scraps collected from industries, but current efforts are directed towards handling mixed and soiled textile waste collected from consumers, which is more difficult to recycle [30]. Some progress has been achieved especially in the case of natural fibers like cellulose-based and protein-based textiles. For example, worn cotton fabrics such as denim are now treated by dissolving the cellulose part using safer chemical solutions like sodium hydroxide and urea. These methods help in recovering usable fiber without adding much harm to the environment. This cellulosic solution is then extruded through a wet spinning process to create new, high quality fibers with mechanical properties comparable to commercial rayon or viscose [31]. Some advanced processes can even preserve the original color of the waste fabric, thereby eliminating the need for a separate, resource intensive re-dyeing step [32]. Regenerated Proteins: Similar principles are being applied to protein-based fibers. Keratin, the protein in wool, can be extracted from discarded wool garments and re-spun into new filaments. Research has demonstrated the ability to produce continuous fibers with a high keratin content (up to 90%) that exhibit properties similar to natural wool [33]. Likewise, casein from waste milk is being investigated as another potential protein source for regenerated fibers [34]. One of important advantages of regenerated fibers, especially those made from cellulose, is their behavior at the end of use. In comparison to synthetic fibers, regenerated cellulosic fibers such as rayon and lyocell break down more easily and fully in natural surroundings like soil and water bodies [35].

2.4 Biodegradable Polymers

With the growing interest in natural and regenerated fibers, researchers have started working on development of bio-based and biodegradable polymers due to their potential to reduce environmental pollution (Table 1). For example, polylactic acid (PLA) is a well-known aliphatic polyester that is produced from renewable plant sources such as corn starch and sugarcane through fermentation [36]. Its industrial production includes ring opening polymerization of lactide with U.S. FDA approval for food-contact applications [37]. It is known to break down under industrial units specific composting conditions (58–80 °C) via hydrolysis of ester bonds, breaking down into CO_2 and water within 6–12 weeks, which provides an option for managing the material after disposal [37, 38]. However, its widespread use is hampered by inherent material limitations, including brittleness at room temperature (strain at break <10%) and poor moisture barrier properties due to high water vapor transmission rate (WVTR), which frequently require modification through plasticizers or blending with other polymers [39, 40]. Polyhydroxyalkanoates (PHAs) are a group of polyesters that are naturally produced by different microorganisms for example by *Cupriavidus necator* as intracellular carbon storage materials [41]. These compounds have recently received more attention because they show better biodegradability in several environments such as soil, freshwater, and marine areas when compared to PLA [42]. Due to this property, PHAs are considered a potential option to reduce the problem of microplastic accumulation, which is commonly seen in products like activewear where fiber shedding often takes place [43]. One of the major areas of confusion among consumers and even in industry is the difference between the terms bio-based and biodegradable. It is important to note that just because a polymer is obtained from a renewable or biological source (bio-based), it does not always mean that the material will break down naturally in the environment [44]. Furthermore, the performance characteristics production costs of many biopolymers (PLA ($11,424/tonne) and PHAs ($2600/tonne) still lag behind their entrenched petroleum-based counterparts such as PET ($1000/tonne) due to energy intensive fermentation, mechanical weakness and thermal instability, presenting significant barriers to their widespread adoption [45, 46]. A common research strategy to overcome these hurdles involves blending biopolymers with natural polymers like starch to improve properties and reduce costs, though this can introduce new challenges related to material compatibility and processing [47]. The changing landscape of textile materials shows that the future of sustainable textiles may not rely on one particular fiber to replace all others. Different types of raw materials are likely to be used together, depending on their specific qualities and uses. Each group of new materials has its own set of benefits, limitations, and suitable applications. For instance, agricultural residue fibers such as coconut fiber are perfectly suited for valorizing local waste streams in agrarian economies, potentially for durable goods like upholstery or composites [48]. Bio-fabricated materials like mycelium offer a high-value, niche aesthetic ideal for the luxury market, where biodegradability and a compelling narrative are key selling points, though their scalability is currently limited [49]. Advanced regenerated cellulose provides a direct, circular solution for the highest-volume waste stream cotton garments but requires significant investment in centralized chemical processing infrastructure [31]. Finally, biodegradable polymers like PHAs are strategically critical for applications with a high risk of environmental leakage, such as performance apparel or disposable nonwovens, directly addressing the persistent problem of microplastic pollution [50]. An intelligent and sustainable resource mix of the future will therefore involve a complex, diversified, and context-aware supply chain, a stark contrast to the current system, which is largely dominated by a few fiber monocultures.

3 Technological Frontiers in Resource-Efficient Manufacturing

Along with the change in raw material sources, important developments are also seen in textile manufacturing processes. Traditional methods, especially wet processing which includes pretreatment, dyeing, printing and finishing, are known to use a large amount of water, energy and chemicals [6]. These steps are considered to be the major contributors to environmental pollution in the textile industry. In recent years, several new techniques have been introduced to reduce the harmful effects of these processes. Some of these include waterless dyeing methods, improved finishing processes, and recycling technologies that help in reusing materials and reducing waste. This section highlights such advancements aimed at making textile production more sustainable.

3.1 The Water Energy Nexus

In textile production, one of the major environmental concerns is the large amount of water used, and energy required to heat it. Among all stages, wet processing

Table 1 Comparative analysis of emerging technologies for sustainable textile fiber production including application in textile industry

Source	Fiber obtained	Properties	Application in textile industry	References
Jute and banana fiber	Cellulose pulp	Chemical: Retained hydroxyl/ carbonyl groups; Mechanical: High crystallinity (XRD verified); Thermal: Stable degradation profile	Packaging reinforcement, regenerated fibers	[51]
Agro-waste biomass	Lignocellulose composites	Structural integrity from lignin-cellulose-hemicellulose matrix; Biological: Biodegradable	Medical dressings, technical textiles	[52]
Wood pulp	Regenerated cellulose	Physical: Tunable pore structure; Mechanical: Enhanced toughness via spinneret geometry optimization	High-performance textiles	[53]
Agricultural by-products	Regenerated cellulosic fibers	Eco-friendly processing; Mechanical: Comparable to commercial lyocell	Apparel, industrial textiles	[54]
Microalgae-ZnO nanoparticles	Functionalized PET/cotton	Biological: Antibacterial (Zn^{2+}-mediated); Chemical: UV-blocking; Mechanical: Retained tensile strength after enzymatic treatment	Smart textiles with anti-UV/ antimicrobial	[55]
Cotton-polyester blends	High-tenacity cellulose	Mechanical: Tenacity = 49.8 ± 1.4 cN/Tex; Thermal: Reversible energy storage	Sustainable clothing, technical textiles	[56]
Waste cotton textiles	Bicomponent cellulose fiber	Color retention from dyed waste; Mechanical: Core–shell structural integrity	Colored textiles without redyeing	[57]
Post-consumer cotton waste	Regenerated cellulose	Compatible with viscose process; Morphological: Uniform fiber formation	Circular fashion products	[58]
Microbial synthesis	PHA derivatives	Thermal: Low melting point (~160 °C); Biological: Marine/soil biodegradable; Mechanical: Ductility comparable to polypropylene	Disposable textiles, medical implants	[46]
Microalgal bioplastics	PHB-rich composites	Hydrolytic stability; Environmental: Degrades within industrial composting timelines	Short-lifecycle apparel	[59]

is considered to be the most water-intensive, contributing nearly 72% of the total water use and around 52% of greenhouse gas emissions from the textile sector [60]. Therefore, there is need to develop new technologies that can significantly reduce or completely remove the use of water in these processing steps.

3.1.1 Supercritical CO_2 ($ScCO_2$) Dyeing

Supercritical fluid dyeing is considered as one of the important developments in textile processing. In this method, carbon dioxide (CO_2) is used after heating and pressurizing it above its critical point (31.1 °C temperature and 7.4 MPa pressure), where it reaches a supercritical state and behaves like both liquid and gas at the same time. In this form, supercritical CO_2 works as a solvent for dyes and replaces the use of water during dyeing [61]. Main benefit of this method is lack of water requirement, which helps in preventing the production of harmful dye wastewater [61]. After the dyeing is completed the pressure is released and CO_2 turns back into gas form that can be collected and reused in the next cycle. Further dye that remains unused settles down as dry powder and can be collected and used again. Furthermore, because the fabric emerges from the process dry, the energy-intensive drying step required in conventional aqueous dyeing is completely omitted, resulting in shorter processing times and lower energy consumption [62]. The technology has proven highly effective for dyeing hydrophobic synthetic fibers like polyester, achieving excellent color strength, uniformity, and fastness properties [62]. However, the widespread adoption of $scCO_2$ dyeing faces two major obstacles. First, the high initial capital investment for the specialized high-pressure equipment is a significant barrier for many manufacturers. Second, the technology effectiveness is largely limited to synthetic fibers [63]. The non-polar nature of $scCO_2$ makes it an excellent solvent for non-polar disperse dyes used for polyester but a poor solvent for the polar, water-soluble dyes required for natural, hydrophilic fibers like cotton and wool [64]. Dyeing natural fibers with $scCO_2$ remains a significant technical challenge, typically requiring complex fiber modifications or the development of entirely new dye chemistries to achieve acceptable results.

3.1.2 Digital and Precision Application

Where water use cannot be completely avoided, the use of precise application methods helps in reducing it to a great extent. Digital Textile Printing (DTP) is now seen as a useful and sustainable option in place of traditional printing methods like rotary and screen printing. Unlike the older methods where large amounts of dye paste are pushed through screens, DTP works like an inkjet printer that sprays small droplets of ink directly on the fabric as per the digital design [65]. This technique avoids the need for screen preparation, helps in reducing chemical waste, use of water (90%) and energy (30–50%) compared to traditional printing. DTP further allows printing as per demand and supports customization, which helps in avoiding the wastage caused by extra or unsold stock [66]. The sustainability and quality of DTP can be further enhanced by pre-treatment process such as atmospheric plasma treatment which modify fiber surface to improve ink adhesion and colorfastness without use of water or harsh chemicals [67]. In addition to DTP, some other dyeing and finishing methods have been developed that help in reducing water use. These include low-liquor or atomization systems. In such technologies, the amount of water needed for dyeing per kilogram of fabric is very low. Methods like foam dyeing and misting work by applying dye and chemicals using stable foam or fine spray, instead of dipping the fabric in large water baths. This allows uniform application on the fabric surface with less water consumption [60]. This approach can reduce water and chemical consumption by as much as 90% in certain applications, such as denim finishing. The primary challenges for these technologies lie in the high initial investment costs for DTP, the need for specialized inks, and, in the case of foam dyeing, the technical difficulty of maintaining foam stability and control in an industrial setting [60].

3.2 Textile to Textile Recycling Technologies

Enabling circular economy in textile sector depends on the ability to recycle used clothes into new textile fibers. Literature highlights two main approaches used for textile to textile recycling: mechanical and chemical recycling (Fig. 1). Mechanical recycling is a widely used method and includes physical processes such as sorting, cleaning, tearing, shredding, and carding of textile waste to recover the fibers. Mechanical recycling method is considered important in reducing textile waste, preventing it from reaching landfill sites and frequently considered as simple and cost effective compared to chemical recycling methods [68]. One of the major advantages of mechanical recycling is that it uses less energy and fewer chemicals, making it more environment friendly [68]. It is a widely used method and can be applied in small scale and localized units. However, there are some limitations such as tearing of fabrics which leads to shortening and damage of fibers which results in reduced fiber strength and quality. As a result, recycled fibers are often used in the production of low-value products like insulation materials, carpet backing, or industrial wipes instead of being reused in high-quality clothing or textiles [29]. While essential for diverting waste from landfill, mechanical recycling, on its own, does not facilitate a long term, high-value circular system for clothing. Chemical recycling is an emerging method that offers better potential for textile circularity. It uses solvents or chemical treatments to break down textile polymers into their basic units such as monomers or purified polymers [69]. For example, glycolysis is used for polyester, and cotton can be treated through dissolution methods. These recovered units can be reused to produce new fibers which are similar in quality to original or virgin materials [70, 71]. This allows the transformation of old clothes into new clothes of the same or better quality, supporting repeated recycling. However, this potential comes with significant challenges. Chemical recycling processes are

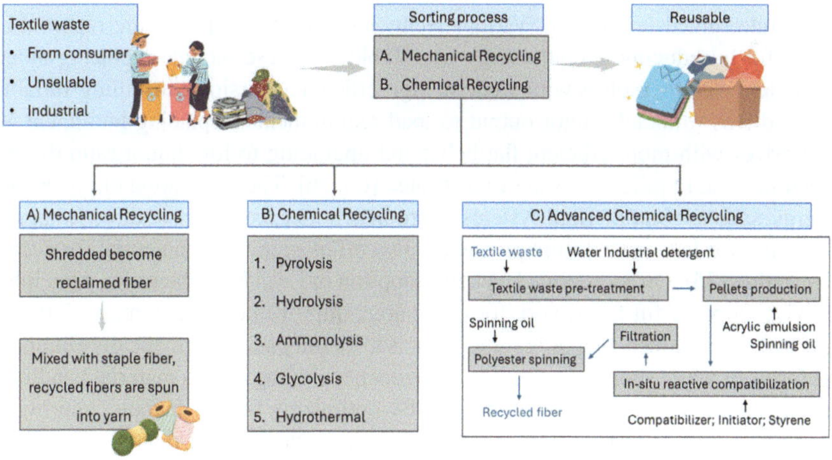

Fig. 1 Textile recycling technologies. **a** Mechanical recycling includes sorting and shredding textiles into reclaimed fibers, which are blended with virgin fibers and spun into yarn. **b** Chemical recycling involves depolymerizing textile waste through methods like pyrolysis, hydrolysis, or glycolysis to recover monomers or petrochemical feedstocks. **c** Advanced chemical recycling uses pelletization and in-situ reactive compatibilization with additives to produce recycled polyester fibers through melt spinning [73, 74]

often highly energy-intensive and can require large volumes of chemicals and water, raising questions about their overall environmental footprint. They are also highly sensitive to feedstock purity; blended fabrics (e.g., cotton/polyester blends), dyes, and finishing chemicals can contaminate the process, making effective sorting and pre-treatment of textile waste an absolute necessity [72]. The economic viability of chemical recycling typically depends on large, centralized, and capital-intensive facilities, which in turn require a consistent, high-volume supply of properly sorted post-consumer textiles a logistical infrastructure that is still in its infancy globally [68].

3.3 Energy Optimization and Lean Manufacturing in Textile Production

Apart from specific innovations in processes, broader strategies are important to improve resource efficiency and ensure sustainability in textile manufacturing. Energy use plays a significant role in textile industries, 40–45% of energy is consumed in manufacturing yarn and fabric, and 35–60% energy is used in the wet processing unit. Major energy consumption takes place in motor-driven units, especially during the spinning process which is highly energy demanding. Steam is also required in large quantity for heating water during wet processing [75]. In addition, humidification and air conditioning systems are used to maintain suitable conditions for uninterrupted production. Various technological improvements are reported to be useful in reducing overall energy use (Table 2). These include replacing older, inefficient electric motors with modern, high-efficiency versions; installing variable speed drives to match motor output to load requirements; replacing inefficient V-belt drives with more efficient flat belts; and upgrading to low-liquor ratio dyeing machines that require less water to be heated [66, 76]. The integration of renewable energy, such as solar thermal systems to pre-heat boiler feedwater or directly supply hot water for low-temperature processes, also offers significant potential for decarbonization [75]. On a systemic level, the adoption of Lean Manufacturing principles provides a powerful framework for continuous improvement. Lean methodologies focus on meticulously mapping value streams to identify and systematically eliminate all forms of waste (non-value-adding activities), including excess material handling, inventory, and energy consumption [77]. By optimizing factory layouts, improving workflows, and fostering a culture of continuous improvement, manufacturers can significantly reduce their operational overhead and overall environmental footprint. Although several studies are available which explain process specific innovations, limited research is found on their integration into a single system. The real challenge for circular economy in textile industry is not only the development of separate technologies, but to link them together into a single, practical, and cost-effective model. For example, the success of a chemical recycling plant for cotton is entirely dependent on the existence of an efficient, large-scale system for collecting and sorting post-consumer textile waste [29]. The recycled fiber produced must then enter a new manufacturing chain, and if it is subsequently dyed using conventional water-intensive methods, a significant portion of the environmental benefit gained through recycling is negated. A truly optimized system would require a chain of innovations: advanced automated sorting, followed by chemical recycling, followed by a waterless dyeing technology like DTP or $scCO_2$ dyeing [29, 63]. Each link in this chain requires immense capital investment, technical expertise, and logistical coordination. The literature often examines these links in isolation, but the most formidable barrier to a sustainable future is forging the connections between them into a functional, integrated whole.

4 Evaluating Sustainability

Increasing use of new materials and production techniques require proper ways to assess their environmental impact and prevent false claims. In some cases, companies make their products look more eco-friendly than they are a practice known as greenwashing where real effects on the environment are not shown clearly [87]. In this section method of LCA is discussed, which play important role to study such environmental claims.

Table 2 Advanced resource efficient processing technologies in textile manufacturing including resources conserved, methodology and outcome

Technology	Resources conserved	Methodology	Applicability/Outcome	References
Digital & precision application				
IoT implementation	Energy, water, raw materials	Sensor networks for real-time monitoring of production lines; Predictive analytics for resource allocation	Optimized production efficiency; Reduced energy/water consumption in global brands (H&M, Nike, Zara)	[78]
Optimization platform	Time, material waste	AI-driven workflow modules for sewing, quality control, and finishing; Synchronization algorithms	Reduced bottlenecks in textile plants; Improved on-time delivery rates	[79]
Deep reinforcement learning	Operational resources	RF models + AHP for multi-criteria decision-making; RL agent for process optimization	Enhanced production scheduling; Adaptability to demand fluctuations	[80]
Knowledge graph construction	Manufacturing resources	Ontology-based NLP for data structuring; Industrial Internet configuration optimization	Standardized resource mapping for network manufacturing	[81]
Textile to Textile Recycling				
Mechanical recycling	Waste diversion, raw fibers	Cutting → Blending → Tearing → Fiber separation → Re-spinning	Applicable to cotton/polyester blends; Output: Fibers for non-apparel industries	[82]
Sustainable design framework	Landfill waste, virgin materials	Lifecycle assessment (LCA) for recyclability scoring; Eco-design principles integration	U.S. context: <15% of 15.4M tons/year textiles recycled; Focus on redesigning recycling-compatible products	[83]
Energy Optimization & Lean Manufacturing				
Lean manufacturing framework	Energy, time, materials	JIT production, 5S organization, Kaizen; Value stream mapping for waste elimination	30% processing time reduction in fishing net manufacturing; Increased service levels	[84]

(continued)

Table 2 (continued)

Technology	Resources conserved	Methodology	Applicability/Outcome	References
Fuzzy optimization model	Grid energy, carbon emissions	Intuitionistic fuzzy sets for solar energy integration; Cost-environment trade-off analysis	Pakistan textile sector; Prioritized solar adoption in energy mix	[85]
Process optimization	Thermal energy, operational costs	Deming cycle (PDCA) for workflow refinement; Bottleneck analysis in dyeing units	Reduced delay risks in shoelace production; Improved machine utilization	[86]

4.1 The Role and Limitations of LCA in a Circular Context

LCA is widely accepted scientific method to measure possible environmental effects of a product or process throughout its full life cycle including all stages from resource collection to production, use, disposal and in some cases, recycling [87]. Various studies compared organic and conventional cotton, chrome based and vegetable based leather tanning, and disposal methods such as recycling and incineration and reported that recycled fibers often show lower environmental impacts compared to those produced from virgin raw materials [88, 89]. It was also observed that type of energy used during production like renewable energy sources or fossil fuel-based electricity plays an important role in influencing the overall environmental footprint, sometimes even more than the production technology itself [90]. In many life cycle assessments, the highest environmental burden is found in parts of the production chain that are not usually expected. For example, in a study on woolen undershirts, the farming of sheep contributed to 88% of the climate change impact, while textile processing showed lesser impact [91]. In case of leather, cattle farming has been found to contribute significantly to environmental emissions, which can reduce the advantages gained through more sustainable tanning practices [92]. While LCA is a useful tool, it also has some challenges, especially when applied to systems that aim to be circular. The method depends heavily on data and the outcomes may vary depending on the study design, system limits, or how by-products are counted because of this comparing different LCA results can be difficult [93]. One of the main concerns is that current LCA models may not fully capture the effects or benefits expected in circular systems. This can lead to a discrepancy between LCA results and consumer or business perceptions of sustainability, potentially enabling circular washing, where a company markets a product as circular based on one attribute (e.g., it contains recycled content) while its overall life cycle impact remains high [94]. In response to these limitations, an emerging theme in the literature is the call for complementing LCA with dedicated circularity indicators that can better account for factors like resource longevity and circular business model potential.

4.2 Conflicting Viewpoints and Ongoing Debates

The idea of textile sustainability involves many discussions and disagreements. These come from different opinions, values, and ways to look at risk and opportunity in the sector. One of the main discussions is about the choice between bio-based and recycled synthetic materials. On one hand, there are materials made from biological sources like polylactic acid (PLA), which can break down naturally but require land, water, and energy for production. On the other hand, there is chemical recycling of synthetic materials such as PET from used plastic bottles. While this helps manage plastic waste, it also keeps the industry dependent on fossil fuel-based products. LCA studies do not show one clear answer, because the better choice depends on which environmental factor is being considered, such as land use, global warming, or toxicity [95, 96]. Is the most sustainable path a highly efficient, large-scale chemical recycling plant that can produce virgin-quality fiber but requires a complex global logistics network to transport sorted waste over long distances? Or is it a network of smaller, decentralized mechanical recycling facilities that produce a lower-quality product but support local economies and have a simpler, lower-impact operational model? This reflects a broader conflict between top-down, capital-intensive solutions and bottom-up, community-scale approaches [97, 98]. Another debate is about large scale technology use versus small scale local systems. While centralized systems may offer higher purity and better output control, decentralized systems allow more flexible and resilient use of local resources [99, 100]. There are also questions about what should drive sustainability in the textile sector government policy or consumer demand. Some believe that rules and policies are needed to push the industry, as seen in the European Union's strategy for circular and sustainable textiles, which includes regulations on design and recycled content [101, 102]. Others think the market should lead, where buyers who care about the environment can support better products. However, many studies show that even though consumers say they support sustainability, their actual buying decisions are still based on cost, ease, and style [103, 104]. Do waterless dyeing and circular materials simply allow the fast fashion model to continue, albeit with a lower impact per item? An alternative perspective, embodied by the slow fashion movement, argues that the true solution lies in sufficiency radically reducing the volume of consumption by emphasizing durability, repair, timeless design, and extending the life of garments [105, 104]. One more major topic is whether new technologies are enough to solve the problems or if people need to change the way they consume. The idea of slow fashion is popular in many studies, but there are doubts about whether it can succeed in a market that depends on fast and low cost products. A lot of research focuses on individual technologies, like a new recycling method or an innovative fiber. But there are not enough studies that connect these technologies into a complete system that can work in real life and meet both environmental and business needs [106, 107].

Research that covers the whole circular process from waste collection to fiber recovery to efficient finishing is limited. These kinds of studies are important to find real problems, check practical use, and measure total impact. Most LCA work in textiles still focuses only on environmental issues. There is very little research on how these new systems and materials affect people and communities, and the use of Social Life Cycle Assessment (S-LCA) is still uncommon [108, 109]. Understanding the social consequences of shifting production models for example, the impact on the livelihoods of cotton farmers if agricultural waste becomes a primary feedstock, or the labor conditions in new bio-fabrication or chemical recycling facilities is critical for ensuring a just and equitable transition [110, 111]. Much of the research on next-generation materials like algae and mycelium focuses on their initial production and properties. There is a lack of long-term studies on their real-world durability, performance through multiple use and wash cycles, and their actual rate of microfiber shedding compared to conventional fibers [112, 113]. Most of the research on sustainable fashion has been carried out in developed countries, mainly in Europe and North America [111]. However, very few studies are available that focus on the practical application of these new technologies and circular business models in the textile-producing countries of the Global South. There is a strong need to develop region specific strategies, keeping in mind the economic limitations, social factors, available infrastructure, and local policies of these nations [114, 115].

5 Conclusion

This book chapter highlights how the textile industry is moving through a phase of change, driven by environmental and social concerns linked with fast fashion practices. Innovations are seen mainly in two areas new material sources and better production methods that use fewer resources. Materials such as algae or fungi based fibers and low water dyeing processes are getting attention as promising alternatives. Still, many of these advances remain at the research or trial stage and are not yet fully applied on a larger scale. There is no single solution that fits all. A sustainable textile future will need a balanced use of different materials and technologies, based on local needs and conditions. For this, the industry must look at the complete system instead of working on isolated changes. Future research should focus on practical models, social impacts, and long-term material performance. Studies should also explore how technologies like AI and blockchain can support tracking and reuse of textiles. Together, these steps can help the industry shift toward a system that respects both the environment and the people involved in textile production.

References

1. Castañeda-Navarrete J, Hauge J, López-Gómez C (2021) COVID-19's impacts on global value chains, as seen in the apparel industry. Dev Policy Rev 39(6):953–970. https://doi.org/10.1111/dpr.12539
2. Shirvanimoghaddam K, Motamed B, Ramakrishna S, Naebe M (2020) Death by waste: fashion and textile circular economy case. Sci Total Environ 718:1–24. https://doi.org/10.1016/j.scitotenv.2020.137317
3. Imran S, Mujtaba MA, Zafar MM, Hussain A, Mehmood A, Farwa UE, Korakianitis T, Kalam MA, Fayaz H, Saleel CA (2023) Assessing the potential of GHG emissions for the textile sector: a baseline study. Heliyon 9(11):e22404. https://doi.org/10.1016/j.heliyon.2023.e22404
4. Dutta P, Rabbi MR, Sufian MA, Mahjebin S (2022) Effects of textile dyeing effluent on the environment and its treatment: a review. Eng Appl Sci Lett 5(1):1–17. https://doi.org/10.30538/psrp-easl2022.0080
5. Uddin F (2021) Environmental hazard in textile dyeing wastewater from local textile industry. Cellulose 28(17):10715–10739. https://doi.org/10.1007/s10570-021-04228-4
6. Petrie L (2023) Sustainability and circularity in the textile value chain a global roadmap. United Nations Environment Programme (UNEP), pp 1–81. www.unep.org
7. Juanga-Labayen JP, Labayen IV, Yuan Q (2022) A review on textile recycling practices and challenges. Textiles 2(1):174–188. https://doi.org/10.3390/textiles2010010

8. Akyildiz SH, Fiore S, Bruno M, Sezgin H, Yalcin-Enis I, Yalcin B, Bellopede R (2024) Release of microplastic fibers from synthetic textiles during household washing. Environ Pollut 357(2024):1–9. https://doi.org/10.1016/j.envpol.2024.124455
9. Bhandari N, Garza-Reyes JA, Rocha-Lona L, Kumar A, Naz F, Joshi R (2022) Barriers to sustainable sourcing in the apparel and fashion luxury industry. Sustain Prod Consum 31(2022):220–235. https://doi.org/10.1016/j.spc.2022.02.007
10. Hajian M, Jangchi Kashani S (2021) Evolution of the concept of sustainability. From Brundtland report to sustainable development goals. In: Sustainable resource management. Elsevier, pp 1–24. https://doi.org/10.1016/b978-0-12-824342-8.00018-3
11. Zaharia RM, Zaharia R (2021) Triple bottom line. In: The Palgrave handbook of corporate social responsibility. Springer International Publishing, pp 1–28. https://doi.org/10.1007/978-3-030-22438-7_2-1
12. Rahman O, Hu D, Fung BCM (2023) A systematic literature review of fashion, sustainability, and consumption using a mixed methods approach. Sustainability 15(16):1–37. https://doi.org/10.3390/su151612213
13. Gautam R (2024) The narrative of circular economy and sustainability -a critical analysis of fashion industry. Circ Econ Sustain 4:3183–3213. https://doi.org/10.1007/s43615-024-00417-y
14. Abbate S, Centobelli P, Cerchione R, Nadeem SP, Riccio E (2024) Sustainability trends and gaps in the textile, apparel and fashion industries. Environ Dev Sustain 26(2):2837–2864. https://doi.org/10.1007/s10668-022-02887-2
15. Jayaprakash K, Osama A, Rajagopal R, Goyette B, Karthikeyan OP (2022) Agriculture waste biomass repurposed into natural fibers: a circular bioeconomy perspective. Bioengineering 9(7):1–12. https://doi.org/10.3390/bioengineering9070296
16. Kozlowski R, Muzyczek M (2023) Hemp, flax and other plant fibres. In: Sustainable fibres for fashion and textile manufacturing. Elsevier, pp 75–93. https://doi.org/10.1016/b978-0-12-824052-6.00017-2
17. Islam T, Rasel SM, Roy R, Hossen MdT, Hossain S, Rahman M, Kabir M, Repon MdR, Maurya SK, Abdul Jalil M (2025) Nature-based flame-retardant finish for cotton fabric using Banana pseudostem saps. Cellulose 32(5):3483–3495. https://doi.org/10.1007/s10570-025-06440-y
18. Pappu A, Patil V, Jain S, Mahindrakar A, Haque R, Thakur VK (2015) Advances in industrial prospective of cellulosic macromolecules enriched banana biofibre resources: a review. Int J Biol Macromol 79:449–458. https://doi.org/10.1016/j.ijbiomac.2015.05.013
19. Akatwijuka O, Gepreel MA-H, Abdel-Mawgood A, Yamamoto M, Saito Y, Hassanin AH (2024) Overview of banana cellulosic fibers: agro-biomass potential, fiber extraction, properties, and sustainable applications. Biomass Convers Biorefinery 14(6):7449–7465. https://doi.org/10.1007/s13399-022-02819-0
20. Wang L, Li G, Chen X, Yang Y, Liew RK, Abo-Dief HM, Lam SS, Sellami R, Peng W, Li H (2024) Extraction strategies for lignin, cellulose, and hemicellulose to obtain valuable products from biomass. Adv Compos Hybrid Mater 7(6):219. https://doi.org/10.1007/s42114-024-01009-y
21. Razali NAM, Mohd Sohaimi R, Othman RNIR, Abdullah N, Demon SZN, Jasmani L, Yunus WMZW, Ya'acob WMHW, Salleh EM, Norizan MN, Halim NA (2022) Comparative study on extraction of cellulose fiber from rice straw waste from chemo-mechanical and pulping method. Polymers 14(3):1–18. https://doi.org/10.3390/polym14030387
22. Prakash S, Radha, Sharma K, Dhumal S, Senapathy M, Deshmukh VP, Kumar S, Madhu, Anitha T, Balamurugan V, Pandiselvam R, Kumar M (2024) Unlocking the potential of cotton stalk as a renewable source of cellulose: a review on advancements and emerging applications. Int J Biol Macromol 261:129456. https://doi.org/10.1016/j.ijbiomac.2024.129456
23. Glova Végsöová O, Straka M (2025) Illegal waste dumps and water quality: environmental and logistical challenges for sustainable development—a case study of the Ružín reservoir (Slovakia). Environments 12(8):251. https://doi.org/10.3390/environments12080251

24. Peng L, Yi J, Yang X, Xie J, Chen C (2023) Development and characterization of mycelium bio-composites by utilization of different agricultural residual byproducts. J Bioresour Bioprod 8(1):78–89. https://doi.org/10.1016/j.jobab.2022.11.005
25. Wattanavichean N, Phanthuwongpakdee J, Koedrith P, Laoratanakul P, Thaithatgoon B, Somrithipol S, Kwantong P, Nuankaew S, Pinruan U, Chuaseeharonnachai C, Boonyuen N (2025) Mycelium-based breakthroughs: exploring commercialization, research, and next-gen possibilities. Circ Econ Sustain 5(2):1–43. https://doi.org/10.1007/s43615-025-00539-x
26. Blanckart L, Munasinghe EA, Bendt E, Rahaman A, Abomohra A, Mahltig B (2025) Algae-based coatings for fully bio-based and colored textile products. Textiles 5(1):1–17. https://doi.org/10.3390/textiles5010003
27. Badruddin IJ, Silva MP, Tonon T, Gomez LD, Rahatekar SS (2024) Modulating the properties of brown alga alginate-based fibers using natural cross-linkers for sustainable textile and fashion applications. ACS Omega 9(35):37002–37011. https://doi.org/10.1021/acsomega.4c03037
28. Almoulkı T, Akkaya E (2023) Using chlorella vulgaris as a natural-textile dye. Environ Res Technol 6(4):326–331. https://doi.org/10.35208/ert.1264338
29. Abtew MA, Atalie D, Dejene BK (2025) Recycling of cotton textile waste: technological process, applications, and sustainability within a circular economy. J Ind Text 55:1–85. https://doi.org/10.1177/15280837251348663
30. Tang KHD (2023) State of the art in textile waste management: a review. Textiles 3(4):454–467. https://doi.org/10.3390/textiles3040027
31. Qiao T, Yang C, Zhao L, Feng Y, Feng X, Mao Z, Wang B (2024) High tensile regenerated cellulose fibers via cyclic freeze-thawing enabled dissolution in phosphoric acid for textile-to-textile recycling of waste cotton fabrics. Int J Biol Macromol 277(1):133911. https://doi.org/10.1016/j.ijbiomac.2024.133911
32. Song R, Fu L, Liu Z, Zhu K, Zhang M, Zhang J, Zeng B, Wang J (2025) Closed-loop recycling for colored cotton textile waste. Sustain Mater Technol 45:e01436. https://doi.org/10.1016/j.susmat.2025.e01436
33. Fang W, Fan R, Aranko AS, Hummel M, Sixta H (2023) Upcycling of keratin wastes in sustainable textile fiber applications. ACS Sustain Chem Eng 11(40):14807–14815. https://doi.org/10.1021/acssuschemeng.3c04987
34. Flores-Nieves MM, Castellanos Espinoza R, Estevez M, Baldenegro-Pérez LA, Trejo JFG, García ME, Cano BM, Soto-Zarazúa GM, España-Sánchez BL (2022) Electrospun Casein fibers obtained from revalued milk with mechanical and antibacterial properties. Arab J Chem 15(11):1–13. https://doi.org/10.1016/j.arabjc.2022.104201
35. Parajuli P, Acharya S, Rumi SS, Hossain MT, Abidi N (2021) Regenerated cellulose in textiles: Rayon, lyocell, modal and other fibres. In: Fundamentals of natural fibres and textiles. Elsevier, pp 87–110. https://doi.org/10.1016/b978-0-12-821483-1.00015-2
36. Shekhar N, Mondal A (2024) Synthesis, properties, environmental degradation, processing, and applications of polylactic acid (PLA): an overview. Polym Bull 81(13):11421–11457. https://doi.org/10.1007/s00289-024-05252-7
37. Goodsel J, Madbouly S (2023) Biodegradable polylactic acid (PLA). Phys Sci Rev 8(7):869–894. https://doi.org/10.1515/psr-2020-0072
38. Greene JP (2021) Bio-based and biodegradable plastics. In: Automotive plastics and composites. Elsevier, pp 149–174. https://doi.org/10.1016/b978-0-12-818008-2.00020-9
39. Kumar S, Graninger G, Falzon BG (2022) Dynamic mechanical thermal analysis (DMTA) of polylactic acid (PLA)/cellulose composite. In: Polylactic acid-based nanocellulose and cellulose composites (1st ed.). CRC Press, pp 165–180. https://doi.org/10.1201/9781003160458-8
40. Sudamrao Getme A, Patel B (2020) A review: bio-fiber's as reinforcement in composites of polylactic acid (PLA). Mater Today: Proc 26(2):2116–2122. https://doi.org/10.1016/j.matpr.2020.02.457
41. Kökpınar Ö, Altun M (2023) Evaluation of different nutrient limitation strategies for the efficient production of poly(hydroxybutyrate-co-hydroxyvalerate) from waste frying oil and

propionic acid in high cell density fermentations of *Cupriavidus necator* H16. Prep Biochem Biotechnol 53(5):532–541. https://doi.org/10.1080/10826068.2022.2114009
42. Tao S, Li T, Li M, Yang S, Shen M, Liu H (2024) Research advances on the toxicity of biodegradable plastics derived micro/nanoplastics in the environment: a review. Sci Total Environ 916:170299. https://doi.org/10.1016/j.scitotenv.2024.170299
43. Garcia Simão RDC, Rocha PMC, Martins JTK, Turkiewicz M, Plewka J, da-Conceição Silva JL, Maller A, Kadowaki MK, Costa-Júnior ÁPS (2024) Exploring biodegradable alternatives: microorganism-mediated plastic degradation and environmental policies for sustainable plastic management. Arch Microbiol 206(12):457. https://doi.org/10.1007/s00203-024-04170-6
44. Terzopoulou Z, Bikiaris DN (2024) Biobased plastics for the transition to a circular economy. Mater Lett 362:136174. https://doi.org/10.1016/j.matlet.2024.136174
45. Nanda S, Patra BR, Patel R, Bakos J, Dalai AK (2022) Innovations in applications and prospects of bioplastics and biopolymers: a review. Environ Chem Lett 20(1):379–395. https://doi.org/10.1007/s10311-021-01334-4
46. Phiri R, Mavinkere Rangappa S, Siengchin S, Oladijo OP, Dhakal HN (2023) Development of sustainable biopolymer-based composites for lightweight applications from agricultural waste biomass: a review. Adv Ind Eng Polym Res 6(4):436–450. https://doi.org/10.1016/j.aiepr.2023.04.004
47. Fredi G, Dorigato A (2024) Compatibilization of biopolymer blends: a review. Adv Ind Eng Polym Res 7(4):373–404. https://doi.org/10.1016/j.aiepr.2023.11.002
48. Hoang T, Van Anh N, Yusuf M, Ali SAM, Subramanian Y, Hoang Nam N, Minh Ky N, Le V, Thi Thanh Huyen N, Abi Bianasari A, Azad KA (2024) Valorization of agriculture residues into value-added products: a comprehensive review of recent studies. Chem Rec 24(8):1–41. https://doi.org/10.1002/tcr.202300333
49. Angelova GV, Brazkova MS, Krastanov AI (2021) Renewable mycelium based composite—sustainable approach for lignocellulose waste recovery and alternative to synthetic materials—a review. Zeitschrift Für Naturforschung C 76(11–12):431–442. https://doi.org/10.1515/znc-2021-0040
50. Acharjee SA, Bharali P, Gogoi B, Sorhie V, Walling B, Alemtoshi (2023) PHA-based bioplastic: a potential alternative to address microplastic pollution. Water Air Soil Pollut 234(1):21. https://doi.org/10.1007/s11270-022-06029-2
51. Rahman MM, Payel MTH, Asaduzzaman M, Hossain S, Ali M (2024) Physical properties of isolated cellulose fiber from jute and banana fiber through kraft pulping: potential applications in packaging and regenerated fibers. SPE Polym 5(4):709–717. https://doi.org/10.1002/pls2.10155
52. Kapoor DU, Gaur M, Kumar A, Singh S, Chidrawar VR, Prajapati BG (2025) Value-added lignocellulose from waste biomass as a potential source in fabrication of lignin-based composite for multifaceted applications. Regen Eng Transl Med 11(2):244–273. https://doi.org/10.1007/s40883-024-00357-x
53. Moriam K, Sawada D, Nieminen K, Ma Y, Rissanen M, Nygren N, Guizani C, Hummel M, Sixta H (2021) Spinneret geometry modulates the mechanical properties of man-made cellulose fibers. Cellulose 28(17):11165–11181. https://doi.org/10.1007/s10570-021-04220-y
54. Plakantonaki S, Stergiou M, Panagiotatos G, Kiskira K, Priniotakis G (2022) Regenerated cellulosic fibers from agricultural waste. AIP Conf Proc 2430:080006. https://doi.org/10.1063/5.0077088
55. Darwesh OM, Al-Balakocy NG, Ghanem A, Matter IA (2023) Application of microalgal-ZnO-NPs for reusing polyester/cotton blended fabric wastes after modification by cellulases enzymes. Waste Dispos Sustain Energy 5(4):471–482. https://doi.org/10.1007/s42768-023-00170-2
56. Tom Mathew B, Bordes R, Samanta A (2025) Waste-derived high-strength cellulosic fibers with reversible thermal energy storage properties. ACS Sustain Chem Eng 2025:1. https://doi.org/10.1021/acssuschemeng.5c03335

57. Rosson L, Byrne N (2022) Bicomponent regenerated cellulose fibres: retaining the colour from waste cotton textiles. Cellulose 29(7):4255–4267. https://doi.org/10.1007/s10570-022-04530-9
58. Zhang X, Zhou W, Wang Z, Xu Y, Zhang G (2025) Manufacturing of post-consumer waste cotton pulp and its regenerated cellulose fiber. Int J Biol Macromol 315:144227. https://doi.org/10.1016/j.ijbiomac.2025.144227
59. Arora Y, Sharma S, Sharma V (2023) Microalgae in bioplastic production: a comprehensive review. Arab J Sci Eng 48(6):7225–7241. https://doi.org/10.1007/s13369-023-07871-0
60. Catarino ML, Sampaio F, Gonçalves AL (2025) Sustainable wet processing technologies for the textile industry: a comprehensive review. Sustainability 17(7):3041. https://doi.org/10.3390/su17073041
61. Abou Elmaaty T, Sofan M, Elsisi H, Kosbar T, Negm E, Hirogaki K, Tabata I, Hori T (2019) Optimization of an eco-friendly dyeing process in both laboratory scale and pilot scale supercritical carbon dioxide unit for polypropylene fabrics with special new disperse dyes. J CO2 Util 33:365–371. https://doi.org/10.1016/j.jcou.2019.06.018
62. Mousa AA, Mohamed FA, Abd El-Megied SA, Youssef YA (2024) Dyeing of synthetic fiber-based wool blended fabrics in supercritical carbon dioxide. Sci Rep 14(1):1–15. https://doi.org/10.1038/s41598-024-81417-8
63. Astha, Neha, Madhu A (2025) Supercritical fluid dyeing technology: innovations and applications. In: Sustainable coloration techniques in textiles. Springer Nature Singapore, pp 237–253. https://doi.org/10.1007/978-981-96-4975-4_8
64. Kreuz A, Horn PA, Sidou LF, Seoud OAE, Andreaus J (2025) Emerging trends in solvent-based textile dyeing. In: Sustainable coloration techniques in textiles. Springer Nature Singapore, pp 113–147. https://doi.org/10.1007/978-981-96-4975-4_4
65. Meena CR, Hada JS (2025) Digital printing and applications in the textiles. In: Advancements in textile coloration. Springer Nature Singapore, pp 349–364. https://doi.org/10.1007/978-981-96-5091-0_13
66. Kumelachew DM, Wagaye BT, Adamu BF (2023) Digital textile printing innovations and the future. In: Digital textile printing. Elsevier, pp 241–259. https://doi.org/10.1016/b978-0-443-15414-0.00006-6
67. Glogar M, Ercegovic Razic S (2024) Sustainable pre-treatment of cellulose knitwear in digital pigment printing processes. Int J Cloth Sci Technol 2024:1–19. https://doi.org/10.1108/ijcst-03-2024-0061
68. Lanz IE, Laborda E, Chaine C, Blecua M (2024) A mapping of textile waste recycling technologies in Europe and Spain. Textiles 4(3):359–390. https://doi.org/10.3390/textiles4030022
69. Harmsen P, Scheffer M, Bos H (2021) Textiles for circular fashion: the logic behind recycling options. Sustainability 13(17):9714. https://doi.org/10.3390/su13179714
70. El Darai T, Ter-Halle A, Blanzat M, Despras G, Sartor V, Bordeau G, Lattes A, Franceschi S, Cassel S, Chouini-Lalanne N, Perez E, Déjugnat C, Garrigues J-C (2024) Chemical recycling of polyester textile wastes: shifting towards sustainability. Green Chem 26(12):6857–6885. https://doi.org/10.1039/d4gc00911h
71. Leenders N, Moerbeek RM, Puijk MJ, Bronkhorst RJA, Bueno Morón J, Van Klink GPM, Gruter G-JM (2025) Polycotton waste textile recycling by sequential hydrolysis and glycolysis. Nat Commun 16(738):1–8. https://doi.org/10.1038/s41467-025-55935-6
72. Ghosh J, Repon MdR, Rupanty NS, Asif TR, Tamjid MI, Reukov V (2025) Chemical valorization of textile waste: advancing sustainable recycling for a circular economy. ACS Omega 10(12):11697–11722. https://doi.org/10.1021/acsomega.4c10616
73. Damayanti D, Wulandari LA, Bagaskoro A, Rianjanu A, Wu H-S (2021) Possibility routes for textile recycling technology. Polymers 13(21):3834. https://doi.org/10.3390/polym13213834
74. Sun G, Cao X, Wang Y, Sun X, Chen Q (2024) Comparative life cycle assessment of two different waste materials for recycled fiber. Resour Conserv Recycl 205:107518. https://doi.org/10.1016/j.resconrec.2024.107518

75. Farhana K, Kadirgama K, Mahamude ASF, Mica MT (2022) Energy consumption, environmental impact, and implementation of renewable energy resources in global textile industries: an overview towards circularity and sustainability. Mater Circ Econ 4(15):1–16. https://doi.org/10.1007/s42824-022-00059-1
76. Özer B, Güven B (2021) Energy efficiency analyses in a Turkish fabric dyeing factory. Energy Sources Part A: Recover Util Environ Eff 43(7):852–874. https://doi.org/10.1080/15567036.2020.1755392
77. Vahabi Nejat S, Avakh Darestani S, Omidvari M, Adibi MA (2022) Evaluation of green lean production in textile industry: a hybrid fuzzy decision-making framework. Environ Sci Pollut Res 29(8):11590–11611. https://doi.org/10.1007/s11356-021-16211-4
78. Abdumajid AA (2024) Application of IOT (internet of things) technologies in the textile industry. Int J Adv Sci Res 05(12):285–289. https://doi.org/10.37547/ijasr-04-12-44
79. Liyanage I, Nuwanga S, Anjana R, Rankothge W, Gamage N (2020) Sustainable manufacturing: application of optimization to textile manufacturing plants. Glob J Comput Sci Technol 20(H2):11–18. https://doi.org/10.34257/gjcsthvol20is2pg11
80. He Z, Tran KP, Thomassey S, Zeng X, Xu J, Haiyi C (2020) A deep reinforcement learning based multi-criteria decision support system for textile manufacturing process optimization, pp 1–18. https://doi.org/10.48550/ARXIV.2012.14794
81. Wang G, Liu G (2021) Construction method of knowledge graph in textile field facing industrial internet. Int Conf Digit Soc Intell Syst (DSInS) 2021:25–29. https://doi.org/10.1109/dsins54396.2021.9670627
82. Nemeša I, Pešić M, Bozoki V (2024) Mechanical recycling of textile waste. Tekstilna Industrija 72(4):24–28. https://doi.org/10.5937/tekstind2404024n
83. Teixeira Franca Alves PH, Bahr G, Clarke-Sather AR, Maurer-Jones MA (2024) Combining flexible and sustainable design principles for evaluating designs: textile recycling application. J Manuf Sci Eng 146(2):020903. https://doi.org/10.1115/1.4063993
84. Carrillo-Corzo A, Tarazona-Gonzales E, Quiroz-Flores J, Viacava-Campos G (2021) Lean process optimization model for improving processing times and increasing service levels using a deming approach in a fishing net textile company. In: Smart innovation, systems and technologies. Springer International Publishing, pp 443–451. https://doi.org/10.1007/978-3-030-75680-2_50
85. Kousar S, Shafqat U, Kausar N, Pamucar D, Karaca Y, Salman MA (2022) Sustainable energy consumption model for textile industry using fully intuitionistic fuzzy optimization approach. Comput Intell Neurosci 2022:1–12. https://doi.org/10.1155/2022/5724825
86. Dao LD, Hanh LD, Minh NX (2024) An application of genetic algorithm into order scheduling of a textile company. Econ Manag 6(24):91–98. https://doi.org/10.59294/HIUJS.VOL.6.2024.634
87. Mousavi K, Kowsari E, Ramakrishna S, Chinnappan A, Gheibi M (2024) A comprehensive review of greenwashing in the textile industry (life cycle assessment, life cycle cost, and eco-labeling). Environ Dev Sustain 2024:1–41. https://doi.org/10.1007/s10668-024-04508-6
88. Avadí A, Marcin M, Biard Y, Renou A, Gourlot J-P, Basset-Mens C (2020) Life cycle assessment of organic and conventional non-Bt cotton products from Mali. Int J Life Cycle Assess 25(4):678–697. https://doi.org/10.1007/s11367-020-01731-x
89. Oliveira M, Zucaro A, Passaro R, Ulgiati S (2024) Life cycle assessment of leather treatment at various scales: comparison between chrome and vegetable processes. Int J Life Cycle Assess 29(2):153–173. https://doi.org/10.1007/s11367-023-02232-3
90. Jhanji Y (2023) Life cycle analysis of textiles and associated carbon emissions. In: Sustainable fibres for fashion and textile manufacturing. Elsevier, pp 403–431. https://doi.org/10.1016/b978-0-12-824052-6.00003-2
91. Bianco I, De Bona A, Zanetti M, Panepinto D (2023) Environmental impacts in the textile sector: a life cycle assessment case study of a woolen undershirt. Sustainability 15(15):11666. https://doi.org/10.3390/su151511666
92. Brugnoli F, Sena K, Zugno L, Oggioni A (2025) A global study on the life cycle assessment (LCA) of the modern cow leather industry. Discov Sustain 6(80):1–21. https://doi.org/10.1007/s43621-025-00798-6

93. Kjaer LL, Pagoropoulos A, Schmidt JH, McAloone TC (2016) Challenges when evaluating product/service-systems through life cycle assessment. J Clean Prod 120:95–104. https://doi.org/10.1016/j.jclepro.2016.01.048
94. Marrucci L, Corcelli F, Daddi T, Iraldo F (2022) Using a life cycle assessment to identify the risk of circular washing in the leather industry. Resour Conserv Recycl 185:106466. https://doi.org/10.1016/j.resconrec.2022.106466
95. Atiwesh G, Mikhael A, Parrish CC, Banoub J, Le T-AT (2021) Environmental impact of bioplastic use: a review. Heliyon 7(9):e07918. https://doi.org/10.1016/j.heliyon.2021.e07918
96. Ramachandran K, Gnanasagaran CL, Vekariya A (2023) Life cycle assessment of carbon fiber and bio-fiber composites prepared via vacuum bagging technique. J Manuf Process 89:124–131. https://doi.org/10.1016/j.jmapro.2023.01.068
97. Huang X, Tan Y, Huang J, Zhu G, Yin R, Tao X, Tian X (2024) Industrialization of open- and closed-loop waste textile recycling towards sustainability: a review. J Clean Prod 436:140676. https://doi.org/10.1016/j.jclepro.2024.140676
98. Kikot M, Malieieva J (2024) Models of forming logistics infrastructure for complex equipment recycling. Innov Technol Sci Solut Ind 3(29):15–28. https://doi.org/10.30837/2522-9818.2024.3.015
99. Chen G, Geng D, Zhu Q (2024) Implementation barriers to the circular supply chain mode for pre-consumer textile waste reprocessing in China. Resour Conserv Recycl 206:107621. https://doi.org/10.1016/j.resconrec.2024.107621
100. Tian X, Peng F, Xie J, Liu Y (2024) Agent-based modeling for an end-of-life power battery cross-regional recycling system and subregional policy analysis: a case study in China. J Clean Prod 441:141054. https://doi.org/10.1016/j.jclepro.2024.141054
101. Charnley F, Cherrington R, Mueller F, Jain A, Nelson C, Wendland S, Ventosa S (2024) Retaining product value in post-consumer textiles: how to scale a closed-loop system. Resour Conserv Recycl 205:107542. https://doi.org/10.1016/j.resconrec.2024.107542
102. Sharif I, Jawaid ST, Khan MN, Siddiqui AH (2024) How does climate policy uncertainty influence sustainable development? Unraveling role of recycling and natural resources in the United States. Miner Econ 37(4):943–960. https://doi.org/10.1007/s13563-024-00475-6
103. Chen X, Tabata T (2024) Circular economy in fashion: consumer awareness and lifestyle, and environmental impact on second-hand clothes. J Mater Cycles Waste Manag 26(6):3876–3892. https://doi.org/10.1007/s10163-024-02091-6
104. Sahimaa O, Miller EM, Halme M, Niinimäki K, Tanner H, Mäkelä M, Rissanen M, Härri A, Hummel M (2024) From simplistic to systemic sustainability in the textile and fashion industry. Circ Econ Sustain 4(2):1115–1131. https://doi.org/10.1007/s43615-023-00322-w
105. Rehman M, Petrillo A, Ortíz-Barrios M, Forcina A, Baffo I, De Felice F (2024) Sustainable fashion: mapping waste streams and life cycle management. J Clean Prod 444:141279. https://doi.org/10.1016/j.jclepro.2024.141279
106. Coscieme L, Manshoven S, Gillabel J, Grossi F, Mortensen LF (2022) A framework of circular business models for fashion and textiles: the role of business-model, technical, and social innovation. Sustain: Sci Pract Policy 18(1):451–462. https://doi.org/10.1080/15487733.2022.2083792
107. Provin AP, Dutra ARDA, De Sousa E, Silva Gouveia ICA, Cubas EALV (2021) Circular economy for fashion industry: use of waste from the food industry for the production of biotextiles. Technol Forecast Soc Chang 169:120858. https://doi.org/10.1016/j.techfore.2021.120858
108. Khan K, Henschel T (2024) LCT-based framework for the assessment of sustainability: from the perspective of literature review. Soc Indic Res 175(3):1–20. https://doi.org/10.1007/s11205-024-03333-8
109. Sigcha E, Sucozhañay D, Cabrera F, Pacheco G, Vanegas P (2024) Applying social life cycle assessment in the informal recycling sector: understanding challenges and limitations. Waste Manag 181:20–33. https://doi.org/10.1016/j.wasman.2024.03.029
110. Ishaq S, Hoang TG, Tanveer U, Hoang T-H, Truong HQ (2024) Transformative capabilities of MedTech organizations in driving circularity in the healthcare industry: insights from multiple cases. J Clean Prod 446:141370. https://doi.org/10.1016/j.jclepro.2024.141370

111. Wagner MM, Heinzel T (2020) Human perceptions of recycled textiles and circular fashion: a systematic literature review. Sustainability 12(24):10599. https://doi.org/10.3390/su12241 0599
112. Ali SS, Al-Tohamy R, Elsamahy T, Sun J (2024) Harnessing recalcitrant lignocellulosic biomass for enhanced biohydrogen production: recent advances, challenges, and future perspective. Biotechnol Adv 72:108344. https://doi.org/10.1016/j.biotechadv.2024.108344
113. Kyriakidis IF, Kladovasilakis N, Pechlivani EM, Tsongas K (2024) Mechanical performance of recycled 3D printed sustainable polymer-based composites: a literature review. J Compos Sci 8(6):215. https://doi.org/10.3390/jcs8060215
114. Abbas A, Zhao C, Ullah W, Ahmad R, Waseem M, Zhu J (2021) Towards sustainable farm production system: a case study of corn farming. Sustainability 13(16):9243. https://doi.org/10.3390/su13169243
115. Kumar R, Gautam, Rani E, Malik A (2025) Cradle-to-factory gate environmental impact of the dairy sector in Haryana, India. Int J Environ Sci Technol 22(3):1735–1750. https://doi.org/10.1007/s13762-024-05747-w

MIX
Papier aus verantwortungsvollen Quellen
Paper from responsible sources
FSC® C105338

If you have any concerns about our products,
you can contact us on
ProductSafety@springernature.com

In case Publisher is established outside the EU,
the EU authorized representative is:
Springer Nature Customer Service Center GmbH
Europaplatz 3, 69115 Heidelberg, Germany

Printed by Libri Plureos GmbH
in Hamburg, Germany